Science and Applications of Conducting Polymers

Science and Applications of Conducting Polymers

Papers from the 6th European Physical Society Industrial Workshop held in Lofthus, Norway, 28–31 May 1990

Edited by

W R Salaneck
Linköping University

D T Clark
ICI Wilton Materials Research Centre

E J Samuelsen
University of Trondheim

CRC Press
Taylor & Francis Group
Boca Raton London New York

CRC Press is an imprint of the
Taylor & Francis Group, an **informa** business

CRC Press
Taylor & Francis Group
6000 Broken Sound Parkway NW, Suite 300
Boca Raton, FL 33487-2742

First issued in paperback 2019

ISBN-13: 978-0-7503-0049-0 (hbk)
ISBN-13: 978-0-367-40300-3 (pbk)

British Library Cataloguing in Publication Data

Science and applications of conducting polymers.
    1. Polymers. Conductivity
    I. Salaneck, W. R.   II. Clark, D. T.   III. Samuelsen, E. J.
    IV. European Physical Society
    620.19204297

Library of Congress Cataloging-in-Publication Data are available

Visit the Taylor & Francis Web site at
http://www.taylorandfrancis.com

and the CRC Press Web site at
http://www.crcpress.com

Preface

This book contains the Proceedings of the 6th Europhysics Industrial Workshop, EIW-6, which focused upon the Science and Technology of Conducting Polymers held at the Hotel Ullensvang, in Lofthus, Norway, during the last week of May 1990. The workshop, initiated by the European Physical Society (EPS) followed the format of previous EPS workshops, in that the attendance was limited to a small number of selected scientists in order to encourage discussion in a relaxed and stimulating atmosphere. Almost 60 delegates attended, representing 14 different countries including Japan and the USA. The speakers were chosen to represent industry, universities and government laboratories.

Although the discovery of conducting polymers dates back essentially only to 1977, the field has grown rapidly over the past ten years. After considering the discoveries, of both scientific and technological interest, which have occurred over the past decade, it was decided that the field is now mature enough that a small conference, a workshop, on the industrial applications of conducting polymers could be useful. In this respect, the EPS industrial workshop in Lofthus was unique.

Topics for presentation orally were chosen on the basis of providing a complete program but with an emphasis upon present and future technological applications. Tutorial lectures, treating the basic physics and chemistry of conjugated polymers, both undoped and doped, to a state of high electrical conductivity (conducting polymers), comprised about one-third of the presentations. Lectures aimed at more technological topics focused upon a wide and interesting array of present and possible future applications including non-linear optical properties and integrated optics, semiconductor devices, sensors, integrated circuit processing, aerospace applications, textiles and high-strength-fiber materials, organic batteries, molecular electronics, and various applications involving both low and high frequency electromagnetic response. The workshop closed with a positive look towards the future role of conjugated polymers, especially within the materials scenario of the chemical industry.

Particularly noteworthy during the industrial discussions were the three following items: An IBM research group reported on the application of radiation-induced doped polyaniline in high spatial resolution electroresists and electron beam discharge layers. Conducting lines as small as 0.25μm have been produced. Lockheed, USA, has recently contracted for the large scale production of polyaniline for use in conjunction with other polymers in the aerospace industry. The foundation of California-based UNIAX, in a joint venture shared by the Neste Corporation, Finland, and a small consortium of prominent American scientists in the field, was also discussed. Several additional industrial representatives alluded to coming announcements about applications of conducting polymers. Very few, however, would allow themselves to be pinned down at this time. More announcements are expected at the International Conference on Synthetic Metals (ICSM-90), Tübingen, Germany, in September 1990.

That the workshop ran smoothly was due to the excellent help of the management and staff of the Hotel Ullensvang, especially Mr Utne and Mr Holm, as well as the administrative organization of Ms Inger Eriksson, Linköping University, and Ms Gillian Cobb, ICI. A special acknowledgment is due to G Thomas, Executive Secretary of the European Physical Society, for his untiring help in all phases of the preparations. In addition, many thanks are due to J Maardalen and S Sunde, Trondheim, for their "guiding light", and to all of the other people who contributed at various stages. Most importantly, we thank the speakers and the authors of posters, and heartily acknowledge a wonderful evening at the wine-and-cheese party hosted by ICI, UK.

The primary sponsor of the workshop was the European Physical Society. We also acknowledge financial support from (in alphabetical order): ICI, UK; IOP Publishing Ltd, UK; Linköping University, Sweden; Neste Oy, Finland; Norwegian Council for Scientific and Industrial Research, NTNF; and the Norwegian Institute of Technology, NTH.

W R Salaneck, D T Clark and E J Samuelsen

Contents

Conducting polymers: The route from fundamental science to technology

Alan. J. Heeger

Institute for Polymers and Organic Solids and Department of Physics, University of
California at Santa Barbara, CA 93106

Abstract: For conducting polymers, a number of areas of potential applications have
developed; each based on new phenomena that are specific to these materials and each
directly traceable to specific aspects of the fundamental science. This "route from
fundamental science to technology" is the focus of this short review. In addition to the
remarkable electrical and mechanical properties which formed the initial (and continuing)
basis of interest in the field, four specific examples are discussed: conducting polymers as
electrochemically active materials, conducting polymers as electrochromic materials,
conducting polymers as nonlinear optical materials, and conducting polymer blends as
composite materials without a percolation threshold.

1. Introduction

With the discovery of conducting polymers in the late 1970's, there was genuine excitement in
the possibility of achieving materials with the important electronic and optical properties of
semiconductors and metals *and* with the attractive mechanical properties and processing
advantages of polymers. Without exception, however, the initial conducting polymer systems
were insoluble, intractable, and non-melting (and thus not processible) with relatively poor
mechanical properties. In addition, the environmental sensitivity of the initial systems proved
to be discouraging.

Remarkable progress has been made. The class of conducting polymers has been greatly
enlarged, and a good underlying understanding of the fundamental molecular features which
are necessary to achieve and control the electronic properties has begun to develop. Soluble
conducting polymers have been discovered which are soluble either in water or in common
organic solvents; solubility in both the conjugated form and in precursor form has been
achieved for a number of polymers (Skotheim,1986; Heeger et al, 1988; and references
therein).

This solubility has enabled processing of films and fibers of conducting polymers and of
blends/composites of conducting polymers with conventional polymers which are co-soluble
in the same solvents. Major improvements have been made in material quality and
environmental stability. Highly oriented materials, in which the macromolecules are chain
extended and chain aligned, have been achieved through post-synthesis tensile drawing
(Tokito et al, 1990; Cao et al, 1990). Measurements carried out on such oriented fibers and
films have demonstrated that conducting polymers can have excellent mechanical properties
and that the mechanical properties and the electrical properties improve together (and in a
correlated manner) as the chain alignment and chain extension are improved.

In parallel with these efforts toward materials improvement, a number of potentially important application areas have been identified. The initial ideas were focused on the possible use of conducting polymers as conductors; i.e. as substitutes for known conductors. Recent progress provides considerable optimism. With electrical conductivities of doped polyacetylene comparable to (or even greater than) that of copper--- and still improving --- this area continues to be exciting. On a weight basis, doped polyacetylene has already been demonstrated to be comparable to or greater than copper (Naarman, 1987; Naarman and Theophilou, 1987; Tsukamoto et al, 1990), and the available data imply that the intrinsic conductivity of chain oriented polyacetylene is still much greater than that achieved to date. Thus the possibility of achieving levels of conductivity in conducting polymers not available from conventional metals appears to be a realistic possibility. Based on current understanding (Kivelson and Heeger, 1988), there is no reason to expect that these remarkable transport properties will be limited to polyacetylene. With comparable materials quality and chain orientation, other conjugated conducting polymers should do just as well.

Equally important is the wide range of electrical conductivity that can be achieved through polymer blends (Andreatta et al, 1988; Hotta et al, 1987). Particularly when conducting polymers are blended with high performance polymers, this wide range of conductivity can be achieved in films and fibers which also exhibit outstanding mechanical properties. Specific examples are fibers of polyaniline blended with poly(p-phenylene terephthalamide) (Andreatta et al, 1990) and fibers of the poly(3-alkylthiophenes) blended with high molecular weight polyethylene (Moulton et al, 1990).

A number of other areas of potential applications have developed; each based on new phenomena that are specific to conducting polymers and each directly traceable to specific aspects of the fundamental science. It is on this "route from fundamental science to technology" that I want to focus in this short review. There are four specific examples: conducting polymers as electrochemically active materials, conducting polymers as electrochromic materials, conducting polymers as nonlinear optical materials, and conducting polymer blends as composite materials without a percolation threshold.

2. CONDUCTING POLYMERS AS ELECTROCHEMICALLY ACTIVE MATERIALS

The novel electrochemistry of conducting polymers was discovered in an attempt to achieve high precision reproducible control over the doping level. This resulted in the discovery that such polymers could reversibly store charge and energy and opened the possibility of high energy density, high power density polymer battery electrodes (for a review see MacDiarmid and Kaner, 1986).

The initial discovery of the ability to dope polyacetylene and control the conductivity over the full range from insulator to metal involved vapor-phase oxidation of the polymer; for example,

$$3/2ny(I_2) + (M)_n \rightarrow [(M)^{+y}(I_3^-)_y]_n \qquad (2.1)$$

where M denotes the monomer unit, such as the CH repeat unit in polyacetylene or the 3-alkylthiophene unit in the P3AT's, etc. Upon exposure to iodine vapor, typically the electrical conductivity increases by many orders of magnitude in just a few minutes. Although such major changes in conductivity are exciting, vapor-phase doping does not allow the kind of control that is required for detailed scientific exploration of the mechanisms involved. It was in direct response to this need for precise control of the doping level that we conceived the electrochemical doping of conducting polymers (Nigrey et al, 1979).

In electrochemical doping, the polymer forms one electrode of an electrochemical cell containing a counterelectrode and an appropriate electrolyte. An example might use lithium metal as the counter-electrode and $LiBF_4$ dissolved in an appropriate solvent as the electrolyte. In such a case, the electrochemical doping reaction would be oxidation of the polymer

$$(M)_n \rightarrow [(M)^{+y}(BF_4^-)_y]_n \qquad (2.2)$$

at the cathode and reduction of the Li^+ ions at the anode, $Li^+ + e \rightarrow Li$ (clearly other electrolytes and counter-ions can be substituted, a reference electrode can be added, etc). In the case of electrochemical doping, the voltage of the external power supply is first set at to the neutral point and then appropriately increased (or decreased) to initiate oxidation (or reduction). The charge injected can be directly measured by integrating the current resulting from the change in electrochemical potential

$$Q = \int i(t)dt \qquad (2.3)$$

from which the total number of charges injected (and thus the doping level) can be obtained, $N = Q/e$. Since control of the external potential allows precise control of the chemical potential of the conducting polymer, since the integration of the charge can be done with high accuracy, and since the steady state with $i(t) \rightarrow 0$ is at electrochemical (and thermodynamic) equilibrium, one can achieve the desired level of control of the doping process.

What was not initially anticipated (and what is so very obvious in hindsight!) was that this process of electrochemical doping opened the field of conducting polymers as electrochemically active materials. This was, nevertheless quickly understood and papers and patents were written documenting this discovery. However, the discovery that conducting polymers were electrochemically active materials was based on the scientific need to precisely and reproducibly control the doping level.

The electrochemical properties of this class of polymers continue to be attractive. Polymer batteries are being developed in a number of countries throughout the world. The unique advantages of materials in which the chemical (and/or electrochemical) potential can be controlled continuously have only begun to be explored.

3. CONDUCTING POLYMERS AS ELECTROCHROMIC MATERIALS

The initial data obtained from electron spin resonance measurements suggested that the changes magnetic properties which occurred on doping were unusual (see the discussion in Heeger et al, 1988). Since the doping occurred without the creation of unpaired spins, there appeared to be a reversal of the spin-charge relations. These initial magnetic observations stimulated the discovery of solitons in polyacetylene, and resulted in the development of a microscopic theory through the SSH model (Su et al, 1979; Su et al, 1980). A principal result of the SSH model was the existence of the mid-gap state in doped <u>trans</u>-polyacetylene; a self-localized electronic state in which the charge is stored and which, by symmetry, appears at the center of the π-π^* gap.

It was immediately clear that the existence of a single mid-gap state was a critical test of the soliton theory and the SSH model, and that such a state could be probed spectroscopically. The schematic band diagram showing the mid-gap state, the interband transition and the doping-induced transition involving the mid-gap state are shown in Figure 1.

The initial confirmation of the existence of the gap state came from spectroscopic measurements carried out on vapor-phase doped samples (Suzuki et al, 1980). However, the advantages of an *in-situ* measurement were immediately obvious; one could carry out the entire set of measurement with a single sample thus enabling quantitative comparison of the spectra at precisely known different doping levels. The results of *in-situ* measurements of the visible-IR absorption, carried out during electrochemical doping, in <u>trans</u>-polyacetylene are

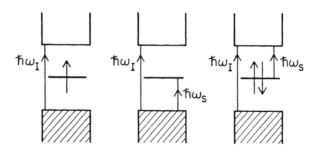

Figure 1. Band diagram showing the mid-gap state associated with the soliton, the interband ($\hbar\omega_I$) and mid-gap ($\hbar\omega_S$) transitions: left, neutral soliton; center, positively charged soliton; right, negatively charged soliton.

shown in Figure 2 (Feldblum et al, 1982). As the doping proceeds, the mid-gap absorption appears, centered near 0.65 - 0.75 eV with an intensity that increases monotonically in proportion to the dopant concentration. At the same time, the strength of the interband transition decreases with an overall conservation of the oscillator strength. From the intensity of this absorption and from the magnitude of the associated bleaching of the interband transition, the width of the self-localized charged object was inferred to be $\xi\approx7a$, in good agreement with the theory.

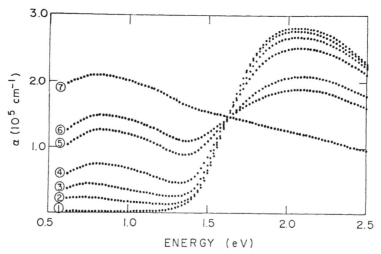

Figure 2.   In-situ absorption curves for trans-$(CH)_x$ during electrochemical doping with the perchlorate ion. The applied cell voltages (relative to Li) and corresponding concentrations are as follows: curve 1, 2.2 V (y = 0); curve 2, 3.28 V (y ≈ 0.003); curve 3, 3.37 V (y ≈ 0.0065); curve 4, 3.46 V (y≈0.012); curve 5, 3.57 V (y ≈ 0.027); curve 6, 3.64 V (y≈ 0.047); curve 7, 3.73 V (y ≈ 0.078). See Feldblum et al, 1982.

The in-situ measurement technique was generalized to other polymers, such as polythiophene (Chung et al, 1984) and polyisothianapththene (Kobayashi et al, 1985) in which the ground state degeneracy had been lifted. The resulting spectroscopic studies provided detailed evidence for the formation of confined soliton pairs (bipolarons) in conducting polymers.

These phenomena and the associated measurement techniques provided the scientific basis for using conducting polymers as electrochromic materials in a number of potential applications ranging from electrochromic displays to "smart" windows which control the transmission and reflection of light. Again, however, the science came first and the realization that the resulting new phenomena had a natural potential for applications came second.

The characteristic spectra and the associated doping-induced changes in absorption found for conducting polymers are sketched in Figure 3; these appear to be general features of conducting polymers (Kobayashi et al 1985). The oscillator strength associated with the interband transition (prior to doping) shifts into the infrared (after doping).

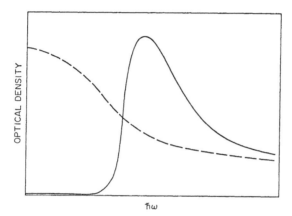

Figure 3. Schematic diagram of the optical density of a conducting polymer as a function of energy. Solid curve: neutral polymer which is transparent for $\hbar\omega < E_g$; dashed curve: heavily doped polymer (metallic).

The effect of such spectral changes depends on the magnitude of the energy gap. If E_g is greater than 3 eV, the undoped insulating polymer is transparent (or lightly colored) whereas after doping the conducting polymer is highly absorbing in the visible. If, however, E_g is small (~1-1.5 eV), the undoped polymer will have a high absorption coefficient (and a relatively high reflectivity) in the visible; whereas after doping the free carrier absorption can be relatively weak provided the typical carrier scattering time (and mean free path) are sufficiently long. Thus the spectral changes lead to electrochromic phenomena; with low voltages characteristic of the electrochemical cell generate doping-induced color changes, changes in transparency and changes in reflectance.

4. CONDUCTING POLYMERS AS NONLINEAR OPTICAL MATERIALS

The large third-order nonlinear optical susceptibility, $\chi^{(3)}$, of conjugated polymers was discovered in the context of the shifts in oscillator strength associated with photogeneration of nonlinear excitations; the mechanism for photogeneration of solitons, polarons, and bipolarons (Su and Schrieffer, 1979). Su and Schrieffer were the first to show that in trans-$(CH)_x$ a photo-injected electron-hole pair should evolve into a pair of solitons within an optical phonon period, or about 10^{-13} seconds. Thus, the absorption spectrum was predicted to shift from $\hbar\omega_I$ to $\hbar\omega_S$ (see Figure 4) on a sub-picosecond time scale. This important

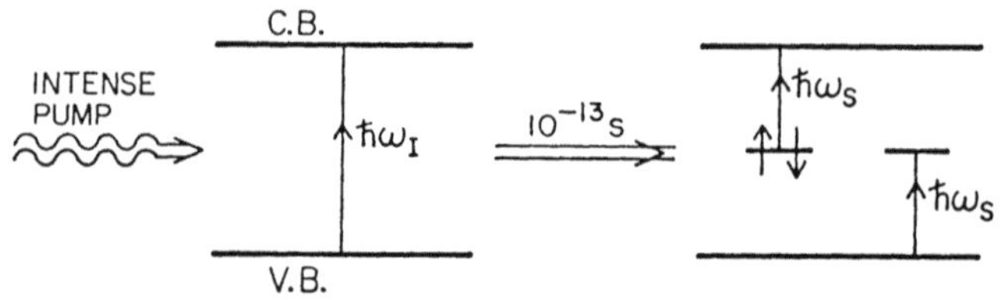

Figure 4. A photopump makes electron-hole pairs, which evolve in 10^{-13}s to soliton pairs with states at midgap. The oscillator strength shifts accordingly. Analogous diagrams can be drawn for the case of polarons or bipolarons.

theoretical work stimulated a major experimental effort using transient pulsed laser spectroscopy (for a summary of results and the relevant references see the review by Heeger et al, 1988). The photoinduced bleaching of the interband transition has been observed on a sub-picosecond time scale; and the reappearance of the oscillator strength in the mid-gap transition has been observed on a sub-picosecond time scale.

Once again, these important and challenging experimental studies were motivated by a basic scientific question: Were solitons photo-generated and if so, on what time scale? The results, however, established conducting polymers as fast-response (sub-picosecond) nonlinear optical materials.

In any material where photoexcitation results in shifts of oscillator strength, the optical properties will be highly nonlinear. For example as a result of the shift in oscillator strength subsequent to photoexcitation, the complex index of refraction is intensity dependent

$$n(\omega) = n_0(\omega) + n_2(\omega,\omega_p)I(\omega_p) \tag{4.1}$$

where the second term describes the nonlinear response at frequent ω to an intense pump at frequency ω_p. The nonlinear index is directly related to the third order nonlinear susceptibility

$$n_2 = 4\pi^2/c\varepsilon\chi^{(3)} \tag{4.2}$$

where ε is the dielectric constant at the probe frequency.

Experiments which pump into the π-π^* transition probe the *resonant* nonlinear response. Although conceptually important for establishing the relevant mechanism responsible for the nonlinearity, resonant pumping is not generally useful in device configurations where absorption of the high pump power could lead to damage. Since the figure of merit, $\chi^{(3)}/\alpha$, is the ratio of $\chi^{(3)}$ to the absorption coefficient (α), nonresonant pumping is the regime of importance.

The Su-Schrieffer mechanism for nonlinear optical response has been generalized into the nonresonant regime; the nonlinear response arises from virtual soliton-antisoliton pairs and is enabled by nonlinear ground state fluctuations, or instantons (Sinclair et al, 1989).

Third harmonic generation (THG) was used to probe the nonlinear susceptibility in oriented samples of polyacetylene Sinclair et al, 1989. The magnitude of $\chi_{\parallel}^{(3)}(3\omega;\omega,\omega,\omega)$ was found to be $(4\pm2)\times10^{-10}$ esu with $\hbar\omega=1.17$eV; the only important component of the third-order susceptibility tensor is that associated with π-electron motion along the backbone. Comparison of THG in cis- and trans-(CH)$_x$ shows that $\chi_{\parallel}^{(3)}|_{trans}$ is 15-20 times larger, implying a mechanism sensitive to the existence of a degenerate ground state. The symmetry specific aspect of $\chi_{\parallel}^{(3)}$ implies a mechanism which is sensitive to the existence of a degenerate ground state; the experimental results indicate that lifting the ground state degeneracy suppresses the nonlinear response. This experimental fact is consistent with a mechanism for $\chi_{\parallel}^{(3)}$ of polyacetylene which involves the generation of virtual solitons enabled by instantons in the ground state. The agreement between the calculated values and the experimental results is remarkable considering that this is a high-order process and that there are no adjustable parameters in the calculation. Thus, the experimental results imply that nonlinear zero point fluctuations lead to an important mechanism for nonlinear optics, particularly in polymers with a degenerate ground state.

The large $\chi^{(3)}$, the ability to withstand relatively high peak pump powers without damage to the sample and the sub-picosecond response have all been demonstrated through detailed experimental studies. Most importantly, the magnitude of $\chi^{(3)}$ is large enough to generate considerable interest in these polymers as nonlinear optical materials.

5. CONDUCTING POLYMER GELS: A SELF ASSEMBLING CONDUCTING NETWORK WITH REMARKABLY LOW PERCOLATION THRESHOLD

The study of connected pathways through percolation theory has provided important insight into the condensed matter physics and materials science of composite systems. Two particular examples are the following:

1) application of percolation theory to the mechanical properties of gels made up from interacting macromolecules dissolved in a solvent;
2) application of percolation theory to the electrical conductivity of composites made up of conducting material in an insulating medium.

In the first case, the system behaves as a complex fluid at polymer volume fractions below the critical percolation threshold. However, at volume fractions sufficiently high that the interacting macromolecules form an infinite connected pathway, a gel is formed in which the connected polymer network becomes rigid and exhibits a finite shear modulus. In the second case, the system behaves as an insulator at volume fractions of the conductor below the critical percolation threshold. However, at volume fractions sufficiently high that the conducting regions form an infinite connected pathway, uninterrupted electronic transport can occur, and the composite exhibits a finite electrical conductivity.

By using soluble conducting polymers, both shear rigidity and electrical conductivity can, in principle, be realized simultaneously with the formation of conducting gels. It is of interest, however, to consider percolation of a conducting polymer on the pre-existing network of a polymer gel. Since the conjugated polymer may adsorb onto the mechanically connected gel network, the formation of conducting paths may be guided by (or assembled by) the pre-existing network. In this way, one can envision obtaining conducting gels at volume fractions of conducting polymer far below that required for percolation (gelation and conductivity) of a soluble conducting polymer alone.

The existence of soluble conducting polymers such as the poly(3-alkylthiophenes) made possible processing of films and fibers of conducting polymers and thereby opened the way to blends/composites of conducting polymers with conventional polymers which are co-soluble in the same solvents.

In a recent study, the UCSB group presented the results of initial measurements on the frequency dependence (from from 1KHz to 1GHz) of the conductivity and dielectric function of conducting gels made by decorating the connected paths of a gel of ultra-high molecular weight polyethylene (≈1-3% UHMW PE in decalin) with conducting polymer (Fizazi et al, 1990). Although the volume fraction (f) of P3OT in the gel ranged from below 0.0005 to about 0.03, there were connected conducting pathways even at volume fractions as low as 0.0005, with no indication of a percolation threshold; i.e. more than three orders of magnitude below the volume fraction for percolation for composite conductors in three dimensions. These unexpected results have led to a deeper fundamental understanding of how to obtain connected conducting pathways in conducting polymer blends. In this example, the materials science involved in the processing of conducting polymer blends generated unexpected progress of a fundamental nature.

Recent studies of conducting gels, made by decorating the connected paths of a gel of ultra-high molecular weight polyethylene (≈1-3% UHMW PE in decalin) with conducting polymer, have demonstrated this phenomenon (Fizazi et al, 1990). Although the volume fraction (f) of poly(3-octylthiophene) in the gel ranged from below 5×10^{-4} to about 3×10^{-2}, connected conducting pathways were found even at $f=5 \times 10^{-4}$, with no indication of a percolation threshold.

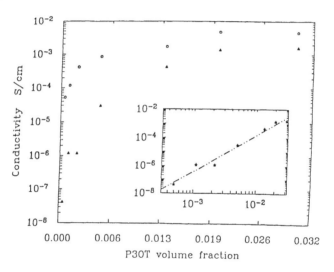

Figure 5: The conductivity is plotted as a function of the volume fraction of conducting polymer for two frequencies, 1kHz and 1GHz. Inset: log-log plot of the conductivity (at 1kHz) vs volume fraction of P3OT.

The percolative aspect of the conducting network is shown in Figure 5 where the conductivity is plotted as a function of the volume fraction of conducting polymer for two frequencies, 1kHz and 1GHz. The 1kHz data characterize the low frequency regime (below ω_c) in which the transport is independent of length scale; the 1 GHz data are in the power law regime (above ω_c) in which the conductivity is strongly sensitive to length scale. Although no specific percolation threshold was observed, for volume fractions of P3OT less than $f \sim 2.5 \times 10^{-4}$ the 1kHz conductivity was too small to measure with the shorted-coax configuration. The conductivity increases by several orders of magnitude for f in the range from $5 \times 10^{-4} < f < 3 \times 10^{-2}$. For f greater than approximately 2×10^{-2}, the conductivity of the network saturates toward values in the range $\sigma \sim 10^{-3}$-10^{-2} S/cm.

The inset to Figure 5 is a log-log plot of the 1kHz conductivity vs volume fraction of P3OT.

The results indicate that over a relatively wide range of concentrations, the conductivity follows a simple power law previously observed for polyaniline/poly-p(phenylene terephthalamide) systems

$$\sigma = c_2 f^\alpha \qquad (5.1)$$

where c_2 is a constant and $\alpha \approx 2.6$. The dashed curve in 4 shows the power law fit to the 1kHz data with $\alpha \approx 2.6$.

Two observations help clarify the nature of the P3OT/UHMW PE gels. First, the conjugated polymer is not expelled from the gel when stored in a larger volume of decalin, even for the lowest concentrations of P3OT. Since decalin is a good solvent for P3OT, this result is important and implies that the conjugated chains are either adsorbed onto or entangled within the PE network. Second, the color of the P3OT/UHMW PE solution changes gradually from yellow to red as the temperature is lowered and gelation occurs. A color change from yellow (the color of P3OT in solution) to red (the color of solid crystalline films) is well-known (Rughooputh et al, 1987) for the poly(3-alkylthiophenes). Reversible, disorder induced thermochromism (red to yellow color change) has also been observed even in solid films (Inganas et al, 1989; Salaneck et al 1989); in this case, it appears that the disorder is due to side-chain melting (Winokur et al, 1989). Thus, the red color of the gel suggests the adsorption of relatively well-ordered P3OT chains onto the PE network, since more complex entanglement of the conjugated polymer would lead to sufficient disorder to cause the P3OT to retain a yellow color.

Percolation theory (Zallen, 1983) predicts that at concentrations sufficiently dilute that there are no connected paths, the conductivity is zero. As f is increased above the percolation threshold (f_p), the conductivity becomes finite and increases as the connectivity (i.e. the number of conducting paths) increases. In contrast, the data presented in Figure 5 show no indication of a well-defined percolation threshold. By attempting to fit the data to the form $\sigma = c_2(f-f_p)^\alpha$, it was concluded that $f_p < 2.5 \times 10^{-4}$ P3OT. By contrast, classical percolation theory (Zallen, 1983) for a three-dimensional network of conducting globular aggregates in an insulating matrix predicts a percolation threshold at a volume fraction $f_p \approx 0.16$, in agreement with results obtained for composites of poly(3-alkylthiophene) in polystyrene (Hotta et al, 1987) and for other conducting polymer composites (Aldissi and Bishop, 1985; Aldissi, 1986).

The extremely low percolation threshhold has, therefore, been understood as arising from decoration of the pre-existing mechanical network of the PE gel by the conducting polymer (Fizazi et al, 1990).

The continuum percolation threshold in two dimensions is inversely proportional to the excluded area (Bug and Safran, 1986). Assuming that the conjugated polymer adsorbs onto the PE network either in the form of rods (length 1) or pancakes (radius of gyration 1), the excluded area in either case is 1^2. The area fraction for percolation, Φ_c, is expressed with a coefficient of order unity, α, as

$$\Phi_c \cong \alpha s / 1^2 \qquad (5.2)$$

where s denotes the actual surface area occupied by the conducting polymer; for a rod $s \cong al$ and for a pancake $s \cong 1^2$, where a denotes the monomer size in the conducting polymer. The volume of the conducting polymer at threshold is expressed in terms of the surface area of the gel network, A,

$$v_c \cong A(\alpha s/1^2)a. \qquad (5.3)$$

The volume of the gel network is expressed in terms of the typical diameter of the structural features (interconnected lamellar crystals, etc; Smith et al, 1981) which make up the network, V=Ad. Thus, the volume ratio of conducting polymer to gel at threshold is $\rho_c = v_c/V$. The critical volume fraction of conducting polymer for percolation in the gel is expressed in terms of the volume fraction of the network in the gel, Φ_0:

$$\Phi_c = \rho_c\Phi_0$$

$$= (a^2/dl)\alpha\Phi_0, \text{ for rods} \qquad (5.4)$$

$$= (a/d)\alpha\Phi_0, \text{ for pancakes.} \qquad (5.5)$$

According to the studies by Smith et al, 1981, d is approximately 100Å. The molecular weight of P3OT and the volume fraction of the PE gel used by Fizazi et al, 1990, are $M_w \approx 89,000$ and $\Phi_0 = 0.02$, respectively. Thus, a≈3Å and l≈1000Å for rods and l≈100Å for pancakes. Thus one obtains $\Phi_c \approx 2 \times 10^{-6}\alpha$ for rods and $\Phi_c \approx 6 \times 10^{-4}\alpha$ for pancakes, consistent with the experimental results.

These relatively highly conducting gels constitute a novel state of matter with an unusual combination of properties. The use of a pre-existing gel network to guide the formation of conducting paths at volume fractions which are orders of magnitude below that needed for bulk percolation may be a useful method for achieving conducting composites with a small volume fraction of conducting polymer.

In fact, this appears to be the mechanism responsible for the absence of a percolation threshold in blends of polyaniline and poly(p phenylene terephthalimide) processed from sulfuric acid (Andreatta et al, 1990). In Figure 6, the conductivity vs volume fraction of

PANI is shown (the inset is a log-log plot); the results indicate that over a relatively wide range of concentrations, the conductivity follows a simple power law as in equation 5.1 where f is the volume fraction of PANI in PPTA. The solid curve in Figure 6 shows the power law fit to the experimental data; the power law dependence holds over a range of values spanning nearly seven orders of magnitude.

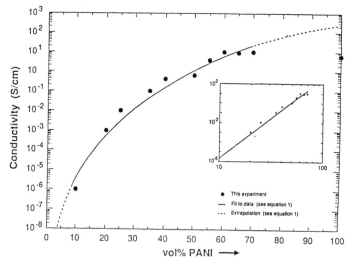

Figure 6. Conductivity versus volume fraction of PANI in the fiber; the inset shows the same data on a log-log plot.

In the PANI/PPTA system, PPTA comes out of sulfuric acid at a water content of about 10% and PANI comes out at a water content of about 25%. Thus, when spinning fibers into a water coagulating bath, the PPTA comes out first, forms a network, and then the PANI adsorbs onto this network in a manner analogous to the P3OT/PE gels. In the PANI/PPTA case, however, this process occurs dynamically during the fiber-spinning process.

6. CONCLUSION

The phenomena responsible for the changes in electrical and optical properties induced by injection of carriers into the π-electron system of conducting polymers (either by doping or by photo-injection) have generated both new concepts and the potential for new applications. In nearly every case, the new "technology" resulted from the exploration of basic scientific questions. In this sense, the evolution of the field of conducting polymers represents a classic case of the route from fundamental science to technology. Based on this excellent progress, there is every reason to believe that these materials will continue to evolve to the point where they can be used in a wide variety of applications.

Acknowledgement: The results described in this review were obtained in collaboration with many students and colleagues. I particularly want to thank Professors Paul Smith, Alan MacDiarmid and Fred Wudl for their many important contributions. I thank the Office of Naval Research (K. Wynne, Program Officer) and the Air Force Office of Scientific Research (D. Ulrich, Program Officer) for support in the specific areas covered in the review.

References:

Aldissi M and Bishop A R 1985 *Polymer* **26** 622
Aldissi M 1986 *Synth. Met.* **13** 87
Andreatta A, Cao Y, Chiang J C, Heeger A J and Smith P 1988 *Synth. Met.* **26** 383
Andreatta A, Heeger A J and Smith P *Polymer* (in press).
Bug A L R, and Safran S 1986 *Phys. Rev. B* **33** 4716
Cao Y, Smith P and Heeger A J *Polymer* (in press)
Chung T C, Kaufman J H, Heeger A J and Wudl F 1984 *Phys. Rev. B* **30** 702
Feldblum A, Kaufman J H, Etemad S, Heeger A J, Chung T C, and MacDiarmid A G 1982
    Phys. Rev. B **26** 815
Fizazi A, Moulton J, Pakbaz K, Rughooputh S D D V, Smith P and Heeger A J 1990
    Phys.Rev.Let. **64** 2180
Heeger A J, Kivelson S, Schrieffer J R and Su W P 1988 *Rev. Mod. Phys.* **60** 782
Hotta S, Rughooputh S D D V and Heeger A J 1987 *Synth. Met.* **22** 79
Inganas O, Gustafsson G and Salaneck W R 1989 *Synth. Met.* **28** C377; and references
    therein.
Kivelson S and Heeger A J 1988 *Synth. Met.* **22** 371
Kobayashi M, Colaneri N, Boysel M, Wudl F, and Heeger A J 1985 *J.Chem.Phys.* **82** 5717
MacDiarmid A G and Kaner R B 1986 *Handbook of Conducting Polymers* ed. T Skotheim
    (New York: Marcel Dekker) Vol. I and Vol. II
Moulton J, Smith P and Heeger A J (to be published)
Naarman H 1987 *Synth. Met.* **17** 223
Naarman H and Theophilou N 1987 *Synth. Met.* **22** 1
Nigrey P J, MacDiarmid A G, and Heeger A J 1979 *J.Chem.Soc.Chem.Commun* 594
Rughooputh S D D V, Hotta S, Heeger A J and Wudl F 1987 *J.Polym.Sci. Polym.Phys.Ed.*
    25 1071
Salaneck W R, Inganas O, Nilsson J O, Osterholm J E, Themans B and Bredas J L 1989
    Synth. Met. **28** C451

Sinclair M, Moses D, McBranch D, Heeger A J, Yu J and Su W P 1989 *Proc. of Nobel Symposium* 73, *Physica Scripta*, **T27** 144

Skotheim T A, 1986 *Handbook of Conducting Polymers* (New York: Marcel Dekker) Vol. I and Vol. II

Smith P, Lemstra P J, Pijpers J P L, and Kiel A M 1981 *Colloid & Polymer Sci.* **259,** 1070

Su W P and Schrieffer J R 1980 *Proc. Nat. Acad, Sci. USA* **77** 5626

Su W P, Schrieffer J R and Heeger A J 1979 *Phys. Rev. Let.* **42** 1698

Su W P, Schrieffer J R and Heeger A J 1980 *Phys. Rev.B* **22** 2099

Suzuki N, Ozaki M, Etemad S, Heeger A J and MacDiarmid A G 1980 *Phys.Rev.Let.* **45** 1209

Suzuki Y, Pincus P and Heeger A J *Macromolecules* (in press).

Tokito S, Smith P and Heeger A J *Polymer* (in press).

Tsukamoto J 1990 *Japanese J. of Appl. Phys .* **29** 1

Winokur M J, Spiegel D, Kim Y, Hotta S and Heeger A J 1989 *Synth. Met.* **28** C419

Zallen R 1983 *The Physics of Amorphous Solids*, (New York: John Wiley) Ch. 4

Theoretical investigation of chain flexibility in polythiophene and polypyrrole

R. Lazzaroni[@], S. Rachidi, J.L. Brédas

Service de Chimie des Matériaux Nouveaux, Département des Matériaux et Procédés, Université de Mons-Hainaut, Place du Parc 20, B-7000 MONS (Belgium)

ABSTRACT: The flexibility of conjugated polymer chains is investigated theoretically for polythiophene and polypyrrole. The possibility for individual rings to rotate around interring single bonds is evaluated by means of quantum chemical calculations, using the ab initio RHF 3-21G and the semiempirical AM1 methods. The rotation barriers in thiophene and pyrrole oligomers are of the order of a few kcal/mole, thereby allowing significant flexibility in solution or in the melt. The introduction of charges on the chain, corresponding to the doping process, leads to a dramatic stiffening of the conjugated backbone, as indicated by a 20-fold increase in the rotation barrier.

1. INTRODUCTION

Strong π-electron conjugation, which gives rise to the outstanding electronic properties of conducting polymers, has long been considered a major obstacle to polymer solubility. This comes as a direct consequence of the extended conformation the backbone takes and of the chain stiffness required to maximize the delocalization of the π wavefunctions. Therefore, synthesizing novel structures with higher conjugation lengths, hence better electrical properties, would push the achievement of solubility even further out of reach. This deadlock was overcome first with the preparation of alkyl-substituted polyaromatics (polythiophene [1-3], polypyrrole [4]) and more recently with the synthesis of alkyl-substituted polyacetylene [5]. The high conformational freedom induced by the side groups is sufficient to provide true solubility of the polymers, when the alkyl chains are at least four carbon-long. This opened the way to detailed investigations of conjugated polymers in solution, namely of the charge transfer processes (doping) in solution and the interplay between the different types of charged defects (polarons, bipolarons), e.g., as a function of solvent [6-8]. The presence of the side groups influences the π-electronic structure of the material by inducing conformational changes in the chain backbone. The solvatochromic and thermochromic effects observed in neutral poly-alkylthiophenes [9,10] are interpreted as the consequences of the rotation of adjacent rings around the interring single bonds. Such torsions lead to a decrease in the π-electron delocalization, and therefore affect the optical properties of the polymer.

In this work, we investigate theoretically the flexibility of polythiophene and polypyrrole chains in both the neutral and the doped states. Using the semiempirical AM1 method [11] combined with ab initio Restricted Hartree-Fock (RHF) 3-21G calculations, we estimate the energy barrier for the rotation of individual rings around the interring single bond in thiophene and pyrrole oligomers of various lengths. From the theoretical values, we deduce the possibility for actual chains to undergo such torsions in solution or in the melt. In the charged systems, the calculated modifications in the molecular structures and in the rotation barriers are related to the experimental data obtained for the doped polymers.

2. THEORETICAL CALCULATIONS

Rotation barriers of the heteroaromatic rings around the interring bond were calculated for the α-bithiophene, α-bipyrrole, and α-quaterpyrrole molecules in the neutral state and in the doubly-charged state, which will be referred to as the positively-doped state. Full geometry optimizations were performed on these systems in the anti-planar configuration using the semiempirical AM1 method. These results were then used as input for ab initio RHF 3-21G calculations of the total energy of different conformations (from anti-planar to perpendicular and finally to syn-planar) in the neutral and "doped" forms. In the tetramer, rotation was considered to take place around the central single bond, between the second and third rings. In the case of pyrrole oligomers, the rotation barrier was also evaluated at the AM1 level; this was not done for bithiophene, since the parameters for the sulfur atom in AM1 are those implemented in the MNDO method, a technique which gives poor estimates of torsion potentials. No further geometry relaxation was allowed during the 3-21G and the AM1 calculations on the different conformers. We have observed that using the rigid rotor approximation, i.e., keeping the geometry of the rings constant while they are rotated, leads only to small differences (less than 10%) in the heigth of the rotation barrier. Even though dimers are short oligomers, we believe they represent valuable models for studying the situation in the polymers. To confirm this hypothesis, we compare the rotation barriers in bipyrrole and quaterpyrrole.

3. RESULTS AND DISCUSSION

a. Bithiophene

-*Neutral state*: The AM1-optimized geometry for anti-planar bithiophene in the neutral state (Fig. 1a) is in good agreement with the experimental structure determined by electron diffraction in the gas phase [12], except for the interring bond which is predicted to be significantly shorter (1.42 Å vs. 1.48 Å).

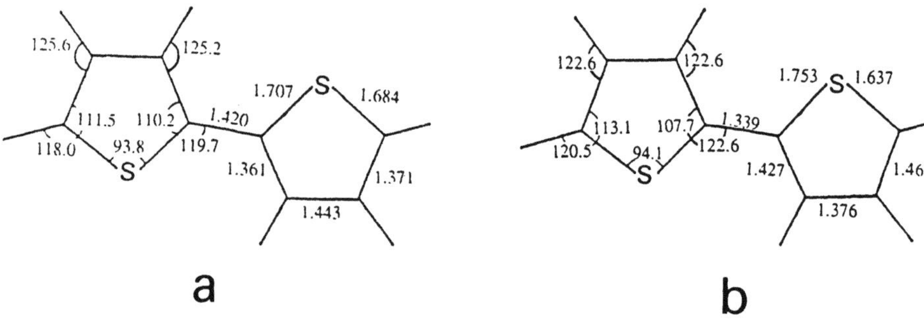

Figure 1: AM1-optimized geometry of the bithiophene molecule in the neutral state (a) and with two positive charges (b). Angles are given in degrees and bond lengths in Å.

As found in previous calculations [13-15], the anti-planar conformation possesses the lowest total energy. Rotating the rings perpendicular to each other induces a 3.2 kcal/mole increase in the total energy at the RHF 3-21G level. This value agrees with the experimental result: 5 ± 2 kcal/mole, obtained from NMR measurements on bithiophene partially oriented in a liquid crystal solvent [16]. The rotation barrier we calculate is intermediate between what was found at the STO-3G level (4.2 kcal/mole [14]) and our previous calculations based on a 3-21G-optimized geometry (2.7 kcal/mole [15]). It is worthwhile to notice that allowing the structure to relax in the perpendicular conformation does not lead to large changes in the bond lengths and bond angles: the largest bond length modification is observed for the interring bond (+ 0.005 Å) and the bond angles vary by less than 0.3°. Consequently, the rotation barrier in the relaxed geometry (3.0 kcal/mole) is very close to the value calculated for the rigid rotor.

Considering the complete loss of π-electron conjugation when rotating the rings to the perpendicular conformation, the increase in total energy remains small. This is explained by the fact that in the perpendicular system no actual π bond is broken and the π orbitals of one ring interact with the σ states of the other one, thereby stabilizing this structure [15,17]. The total energy of the syn-planar conformation is 1.7 kcal/mole higher than that of the anti-planar form, probably because of small steric hindrance between the hydrogen atoms linked to the β carbons. Calculations on intermediate conformations show that the system where the rings are twisted 30° away from the anti conformation possesses a total energy only 0.3 kcal/mole higher than the fully planar one. Such a small increase indicates that significant deviations from planarity are definitely allowed at room temperature. Macroscopically, this flexibility translates into the well-known solvatochromic and thermochromic effects observed in polyalkylthiophenes. In the former phenomenon, interaction with a good solvent leads to chain twisting. The persistence length of poly-3-butylthiophene in chloroform solution, measured by small angle neutron scattering, is 55 Å [18]. This means that besides torsions around the anti-planar conformation, syn segments, which reduce the elongation of the chain, probably exist in solution. On the other hand, thermochromism arises above the "melting point" of the alkyl chains; the conformational freedom acquired by the side groups is then partly transmitted to the polymer backbone. In both cases, ring torsions decrease the π conjugation sufficiently to induce changes in the energy of the lowest π-π* transition.

-Doped state: The presence of two positive charges on the bithiophene molecule strongly modifies its geometry towards a quinoid configuration (Fig. 1b; anti-planar conformation), as already pointed out in previous studies [19,20]. Compared to the neutral state, the "inside" and "outside" Cα-Cβ bonds are 0.095 and 0.065 Å longer, respectively, while the Cβ-Cβ bonds shorten by 0.067 Å. More importantly, the interring bond length decreases by 0.081 Å to 1.339 Å, thereby acquiring a complete double-bond character. Rotating the rings in the perpendicular conformation does not induce large changes in the inner geometry of the rings; in contrast, the interring bond elongates to 1.379 Å. While this represents a marked change from the anti-planar geometry, it is still far from the interring distance of the neutral molecule (1.420 Å), indicating that strong electronic interaction between the rings is preserved even in this situation.

Within the rigid rotor approximation, the RHF 3-21G total energy of the perpendicular system is 75 kcal/mole higher than that of the anti-planar conformation, whereas the syn-planar form is only slightly more energetic (ΔEsyn-anti = 1 kcal/mole). At this point, it must be stressed that the 75 kcal/mole value must be considered only as a rough estimate of the rotation barrier in this case. This is due to the facts that: i) in principle, the perpendicular conformation allows for the presence of a biradical configuration that RHF calculations do not describe accurately, and ii) the significant relaxation of the interring bond mentioned above actually stabilizes the perpendicular system, lowering the barrier. In fact, we remained within the rigid rotor approach in order to build the potential curve displayed on Figure 2. The dashed part of the curve represents the torsion range in which the above-mentioned limitations may play an important role.

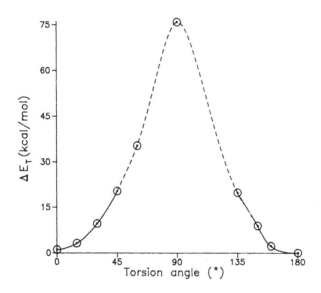

Figure 2: RHF 3-21G-calculated potential curve for doped bithiophene,
as a function of the angle between the thiophene rings
(180° corresponds to the anti planar conformation).

In any case, the definite message from these results is that the rotation barrier in the doped state is much higher than in the neutral state. As a consequence, the total energy grows very steeply when moving away from planarity. For instance, a 30°-deviation from the anti-planar conformation requires $\approx$ 9 kcal/mole in the doped state, while only 0.3 kcal/mole are needed for the same torsion in the neutral molecule. Therefore, in the doped polymer, the chains have to remain completely planar, hence they are much stiffer than in the undoped state. This evolution has been observed experimentally as the persistence length of poly-3-butylthiophene in chloroform increases from 55 Å to over 800 Å upon doping [18]. Another consequence of our calculations is to suggest that polythiophene segments "caught" in the syn-planar conformation (which is accessible from the anti-planar form in the neutral state) when the charge transfer occurs would be trapped in that configuration due to the huge potential barrier in the doped state. Finally, we note that the electronic properties calculated for the syn- and anti-planar conformations are almost identical, be it in the undoped or doped state.

b. Pyrrole oligomers

-Neutral state: The optimized geometries of bipyrrole and quaterpyrrole in the anti-planar conformation (Fig. 3a,c) are in good agreement with the X-ray data obtained on terpyrrole [21].

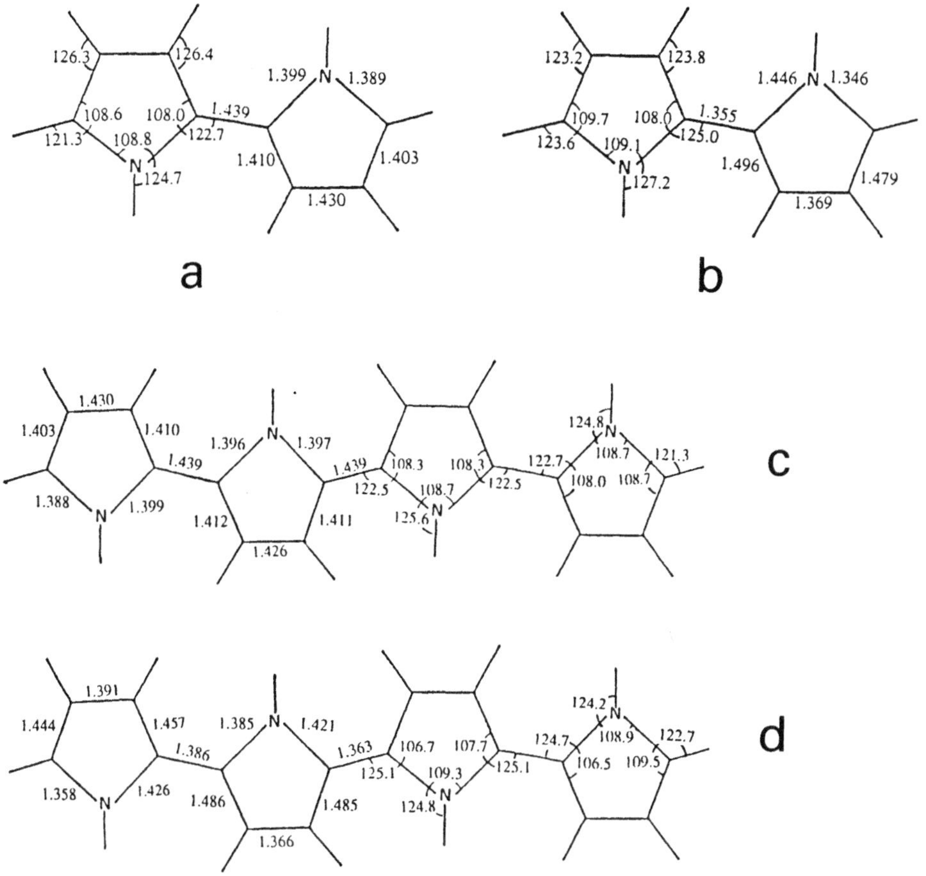

Figure 3: AM1-optimized geometry of bipyrrole and quaterpyrrole in the neutral state (a,c) and with two positive charges (b,d). Angles are given in degrees and bond lengths in Å.

No large modifications in bond lengths and bond angles are observed as the oligomer gets longer; the geometry of the inner part of the bipyrrole molecule is kept unchanged along the quaterpyrrole backbone, except for the outer $C\alpha$-$C\beta$ and $C\alpha$-N bonds. Compared to bithiophene, the degree of bond length alternation between the $C\alpha$-$C\beta$ and $C\beta$-$C\beta$ bonds is much less pronounced (≈ 0.02 Å vs. ≈ 0.08 Å), reflecting the higher aromatic character of the pyrrole rings. Note that the same trend is found in the experimental geometries [12,21]. The interring bond is calculated to be slightly longer than in bithiophene (1.439 Å vs. 1.420 Å); this value is in good agreement with the X-ray diffraction result (1.448 Å), in contrast to what was observed in bithiophene.

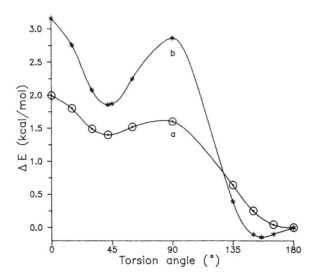

Figure 4: AM1 (a) and RHF 3-21G (b) potential curves for neutral bipyrrole as a function of the torsion angle between the pyrrole rings. The vertical axis represents the total energy in 3-21G and the heat of formation in AM1.

Figure 4 shows the potential curves for the torsion between the rings in bipyrrole calculated with the AM1 and RHF 3-21G methods, using the rigid rotor approximation. It appears clearly that the overall shape of the torsion potential is similar with both techniques. At the AM1 level, the curve minimum corresponds to the anti-planar form, whereas the 3-21G calculations yield a minimum for a 155°-twisted conformation. However, in this case, the difference in total energy relative to the anti-planar system is only 0.1 kcal/mole, indicating that the planar conformation is certainly allowed at room temperature. This reconciles the RHF 3-21G results with the fact that terpyrrole is found to be planar in the solid state [21].

As in bithiophene, the perpendicular conformation corresponds to a maximum in the potential curve; in bipyrrole, the rotation barrier is calculated to be 2.9 kcal/mole at the RHF 3-21G level, i.e., very close to the value in bithiophene, and 1.6 kcal/mole at the AM1 level. Comparison between the two curves indicates that the overall shape of the potential is well estimated with AM1, but this technique provides too low a value for the rotation barrier. Nevertheless, these results demonstrate clearly that strong deviations from planarity are allowed in bipyrrole at room temperature: the total energy of the 135°-conformation is only 0.5 kcal/mole higher than the curve minimum. Again, this points to the possibility for long polymer chains, here polypyrrole, to undergo significant twisting in solution.

When going over the perpendicular conformation towards the syn-planar form, the potential exhibits a well-defined local miminum around 40°; at the RHF 3-21G level, the energy barrier between this conformation and the perpendicular structure is ≈ 1 kcal/mole. The existence of this second minimum is due to the fact that the syn-planar form is strongly destabilized in bipyrrole, which is not the case in bithiophene. This conformation corresponds to the maximum in the potential curve, with a total energy even higher than that of the perpendicular form. Such a destabilization is a consequence of the strong steric hindrance between the two hydrogen atoms linked to the nitrogens. In the totally planar conformation, the distance between these two hydrogens is only 2.19 Å, i.e., less than the double Van der Waals radius (1.2 Å). Obviously, this phenomenon does not take place in thiophene oligomers. Therefore, the major difference between the torsion behaviour of bipyrrole and bithiophene is the possibility for a stable syn-gauche conformation to exist in the former molecule, whereas in the latter, only the two planar conformations are expected to exist.

The potential curve calculated for quaterpyrrole at the AM1 level (this system is too large to allow for 3-21G calculations) is almost identical to that determined in bipyrrole: the rotation barrier is 1.5 kcal/mole (vs. 1.6 kcal/mole in the dimer), ΔEsyn-anti is 1.7 kcal/mole (2.0 kcal/mole in the dimer), and the local minimum is located around 40°. This similarity indicates that: i) the trends observed for the dimers are preserved quantitatively in longer oligomers and probably in the polymers, and ii) the existence of a stable syn-gauche conformation in pyrrole oligomers is likely to give rise to helical conformations in polypyrrole.

-Doped state (Fig. 3b,d): The evolution of the geometry of pyrrole oligomers upon "doping", i.e., when introducing two positive charges on the molecule, is parallel to what is described for bithiophene: we observe a reversal of single- and double-bond character inside the rings and for the interring bonds. In quaterpyrrole, the major structural changes take place in the two central rings, even though the positive charge spreads equally over the four pyrrole units. The interring bond shortens to 1.355 Å and to 1.363 Å, in the dimer and the tetramer (central bond), respectively. The decrease in the length of this bond relative to the neutral state (0.084 Å in the dimer and 0.076 Å in the tetramer) is very close to the value for bithiophene (0.080 Å).

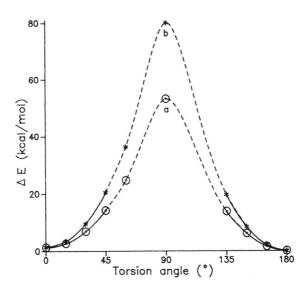

Figure 5: AM1 (a) and RHF 3-21G (b) potential curves for doped bipyrrole as a function of the torsion angle between the pyrrole rings. The vertical axis represents the total energy in 3-21G and the heat of formation in AM1.

Consistently, the rotation barrier around the interring bond is also very high ($\approx$ 80 kcal/mol at the 3-21G level; Fig. 5), indicating again the stiffness of the molecule in the doped state. As observed previously, the AM1 calculations provide a good estimate of the overall shape of the potential curve, even though the value of the barrier is significantly lower: approximately 60% of the result obtained with the abinitio technique. As in the case of bithiophene, the absolute value of the total energy for the perpendicular conformation has to be taken with some precaution. Nevertheless, the rotation barrier in the doped molecule is far higher than in the neutral state.

It is noteworthy that the local minimum corresponding to the syn-gauche form has disappeared upon doping; the syn-planar conformation is 1.5 kcal/mole less stable than the anti-planar system in the RHF 3-21G calculations. The smaller destabilization, compared to the neutral state (ΔEsyn-anti = 3.1 kcal/mole), may be a consequence of the fact that the doping induces slight increases in the N-Cα-Cβ and H-N-Cα angles, by 2.7° and 2.5°, respectively. Therefore, the hydrogen atoms linked to nitrogen are moved away from each other and are now located 2.41 Å apart, i.e., right above the double Van der Waals radius for hydrogen. Finally, the AM1-calculated potential curve for doped quaterpyrrole is very similar to the curve determined for the dimer: the rotation barrier is $\approx$ 45 kcal/mole and the two minimums correspond to the planar situations. Again, this demonstrates that the trends observed in the dimers remain valid in longer oligomers.

4. SYNOPSIS

The possibility of ring torsions in conjugated thiophene and pyrrole chains in the neutral state is clearly demonstrated by the results of our calculations: in bithiophene and bipyrrole, the 3-21G-determined rotation barrier is $\approx$ 3.0 kcal/mole, and deviations from the anti-planar form well above 30° are allowed at room temperature. Bipyrrole is shown to possess a stable syn-gauche conformer, which is preserved in longer oligomers and most probably in the polymer. Therefore, polythiophene and polypyrrole chains are flexible enough to adopt non planar and, in the case of polypyrrole, helical conformations in solution or in the melt. In sharp contrast, the dramatic change in the chain geometry upon charge transfer makes the backbone of the doped systems much stiffer, as the interring bond acquires a strong double-bond character. The barrier for rotation around this bond is much higher than in the neutral case, thereby preventing any significant ring rotation and chain twisting. This evolution is in agreement with the small angle neutron scattering results on polymer solutions.

ACKNOWLEDGEMENTS

R.L. is indebted to the Belgian National Fund for Scientific Research (FNRS) for financial support. We are grateful to the University of Mons for the use of the UMH-CCI Computer Center as well as FNRS and IBM-Belgium for the use of the supercomputer network at KU Leuven, U. Liège, and FNDP Namur.

REFERENCES

@ Chargé de Recherches du Fonds National de la Recherche Scientifique (FNRS).

1. R.L. Elsenbaumer, K.Y. Jen, and R. Oboodi; Synth. Met. 15, 169 (1986).
2. M. Sato, S. Tanaka, and K. Kaeriyama; J. Chem. Soc. Chem. Commun. 873 (1986).
3. S. Hotta, S. Rughooputh, A.J. Heeger, and F. Wudl; Macromolecules 20, 212 (1987).
4. E.E. Havinga, L.W. van Horssen, W. ten Hoeve, H. Wynberg, and E.W. Meijer; Polym. Bull. 18, 277 (1987).
5. E.J. Ginsburg, C.B. Gorman, R.H. Grubbs, F.L. Klavetter, N.S. Lewis, S.R. Marder, J.W. Perry, and M.J. Sailor; in *"Conjugated Polymeric Materials: Opportunities in Electronics, Optoelectronics, and Molecular Electronics"*, J.L. Brédas and R.R. Chance Eds., NATO ASI Series 182 (Kluwer, Dordrecht, 1990).
6. M.J. Novak, S. Rughooputh, S. Hotta, and A.J. Heeger; Macromolecules 20, 965 (1987).
7. D. Spiegel, P. Pincus, and A.J. Heeger; Synth. Met. 28, C385 (1989).
8. M.J. Novak, D. Spiegel, S. Hotta, A.J. Heeger, and P. Pincus; Macromolecules 22, 2917 (1989).
9. S. Rughooputh, S. Hotta, A.J. Heeger, and F. Wudl; J. Polym. Sci. Polym. Phys. 25, 1071 (1987).
10. W.R. Salaneck, O. Inganäs, B. Thémans, J.O. Nilsson, B. Sjögren, J.E. Österholm, J.L. Brédas, and S. Svensson; J. Chem. Phys. 89, 4613 (1988).
11. M.J.S. Dewar, E.G. Zoebisch, E.F. Healy, and J.F. Stewart; J. Am. Chem. Soc. 107, 3092 (1985).
12. A. Almenningen, O. Bastiansen, and P. Svendsås; Acta Chem. Scand. 12, 1671 (1958).
13. J.L. Brédas, G.B. Street, B. Thémans, and J.M. André; J. Chem. Phys. 83, 1323 (1985).
14. V. Barone, F. Lejl, N. Russo, and M. Toscano; J. Chem. Soc. Perkin Trans. II, 907 (1986).
15. J.L. Brédas and A.J. Heeger; Macromolecules 23, 1150 (1990).
16. P. Bucci, M. Longeri, C.A. Veracini, and L. Lunazzi; J. Am. Chem. Soc. 96, 1305 (1974).
17. E. Ortì, J. Sanchez-Marin, and F. Tomas; J. Molec. Struct. Theochem 108, 199 (1984).
18. J.P. Aimé, F. Bargain, M. Schott, H. Eckhardt, G.G. Miller, and R.L. Elsenbaumer; Phys. Rev. Lett. 62, 55 (1989).
19. J.L. Brédas, B. Thémans, J.G. Fripiat, J.M. André, and R.R. Chance; Phys. Rev. B 29, 6761 (1984).
20. S. Stafström and J.L. Brédas; Phys. Rev. B 38, 4180 (1988).
21. G.B. Street; in *"Handbook of Conducting Polymers"*, T.A. Skotheim Ed. (Dekker, New York, 1986).

Conjugated and non-conjugated polymers in integrated optics

D. Bloor

Applied Physics Group, School of Engineering and Applied Science, University
of Durham, South Road, Durham, DH1 3LE, U.K.

ABSTRACT: The use of conjugated and non-conjugated polymers in active and
passive integrated optic waveguides and devices is reviewed. Optical
waveguides can be fabricated from commercially available polymers using
simple processing methods. Examples of the structures that can be produced
are presented. The use of these for thermal sensors is described. Progress
towards active devices employing polymers with non-linear optical pendent
groups offers prospects for competitive electro-optic modulators and
switches. The use of conjugated polymers for all optical devices requires
further advances in materials and processing.

1. INTRODUCTION

The inroads made by fibre optics into the communications, data transmission
and interconnect fields has resulted in a need for improved devices for
optical signal processing. The direct modulation, amplification and pre-
processing of optical signals obviates the need for additional optical to
electronic conversion, and vice versa, in systems. Progress in this direction
requires both new materials, with large non-linear optical (nlo) coefficients,
good stability and ease of processing, and the development of integrated
optical devices, which must be compatible with fibre optics, show reductions
in size and power consumption and operate at higher frequencies.

Organic compounds have been shown to possess large nlo coefficients but are
difficult to prepare as single crystals with dimensions suitable for optical
waveguides, see for example Chemla and Zyss (1987), Zyss (1985) and Hann and
Bloor (1989). In contrast polymers can be readily prepared in fibre and thin
film form and offer low optical loss, c.f. Kaino (1986) and references
therein. The properties of optical waveguides also enable the structures to
be designed to optimise nlo effects, e.g. second harmonic generation, Zyss
(1985). There has in consequence been considerable interest in polymers with
either nlo active pendent groups or conjugated backbones, for electro-optical
and all optical devices respectively, Prasad and Ulrich (1988) and Heeger et
al., (1988).

Spin and dip-coating are simple routes providing large area thin films of
polymers suitable for planar waveguides. Device construction requires, in
addition, a means of producing a confining channel. This can be provided by
modifying the refractive index of either the polymeric film or the substrate,
by physically defining the channel as a rib in the polymer or a channel in the
substrate or finally by depositing a dielectric strip on top of the polymer

film.   It is this ease and flexibility of fabrication which makes polymers an attractive medium for integrated optics.

Examples are presented below of the use of nlo-inactive commercially available polymers, and nlo-active research polymers in the fabrication of both passive and active integrated optic structures.

2.   WAVEGUIDES IN COMMERCIAL POLYMERS

2.1 Passive waveguides

The success in producing polymeric optical fibres from polymethyl methacrylate (PMMA) and deuterated and fluorinated analogues, Kaino (1986), has resulted in the investigation of the suitability of a number of commercially available polymers for planar integrated optics.

Table 1.   The properties of polymeric planar optical waveguides measured at 632.8nm

Polymer	n(TE)	n(TM)	Loss (dB/cm)	Reference
PMMA	1.508 to 1.542		1.8 - 4.2	Driemeier and Brockmeyer (1986)
PMMA	1.494 to 1.520		0.3 - 3	Okamoto and Tashiro (1988)
PMMA	-	-	0.1 - 1	Schriever et al. (1985)
Polyurethane	1.593 1.595	- -	10 - 20[a]) 6 - 12[b])	Kapoor et al. (1989)
Polyurethane	1.56	1.56	< 1	Diemeer et al.(1989)
Polyimide	1.690	1.645	~ 10	Savatinova et al. (1990)
Polyimide	1.59	-	> 1.2	Burzynski et al. (1988)
Polyimide	1.641	1.638	~ 1	Franke et al. (1986)
Polyvinylpyridine	1.572	1.572	1.6	Wells (1990)
Epoxy resin[c]	1.54	1.54	0.25 - 0.5	Hartman et al. (1989)
Polyguide	1.58	1.58	0.25 (830 nm)	Brown (1989)
Plex 6696.0	1.513	1.513	0.25	Driemeier (1990)

a.   Polyester-polyurethane
b.   Acrylic-polyurethane
c.   Large multimoded guides

Selected examples from the literature are listed in Table 1. Refractive indices are quoted for TE and TM modes. The values indicate the film birefringence for the electric vector normal to and in the plane of the polymer film. Optical guiding has been used previously to determine the birefringence, which was attributed to preferential molecular alignment, Prest and Luca (1979). However, uneven strain produced during film casting can also produce birefringent films, Reuter et al., (1988). The literature data in Table 1 shows examples with both zero and large birefringence. There are also large variations in refractive index for a given class of polymer. The latter differences reflect gross effects, e.g. different chemical composition within a class of polymers, the presence of additives, etc., while the former are probably due to more subtle differences between samples, e.g. molecular weight distribution, degree of crosslinking, etc. To date little attention has been paid to these details. However, the much lower losses for carefully prepared PMMA fibres, typically less than 10^{-2} dB/cm, suggests that a more thorough approach to the preparation of planar waveguides will result in better performance than that shown in Table 1.

Most work has been limited to studies of single polymer layers which act as planar guides. However, a more extensive study has been made of waveguide structures using poly-4-vinylpyridine, (P4VP), Wells (1990). Thin films of P4VP were formed by dip-coating on soda and silica glass and silicon with a silicon oxynitride buffer layer. The thickness of these films was controlled by variation in the concentration of the P4VP-isopropylalcohol solution and the withdrawal rate. Film thickness was determined by a mechanical profilemeter, interference fringes recorded at normal incidence and from the waveguide mode structure. Agreement between these methods for films of thickness 1 μm was better than 10%. The dependence of film thickness on withdrawal rate shown in Figure 1 is close to that predicted theoretically, Yang et al., (1980). Waveguide properties of the films were determined by the prism coupling technique, Stegeman et al., (1987). Analysis of the TE and TM waveguide mode structure gave a refractive index close to 1.57 with little birefringence for films prepared using P4VP free from crosslinks. The birefringence was found to be a function of film thickness and mode number, see Figure 2. This and the observation of strong birefringence in films made from polymer containing crosslinks suggests strongly that the birefringence is due to anisotropic strain. P4VP films deposited on single crystal silicon with a silicon oxynitride buffer layer gave clean end-faces on substrate cleavage. Far-field mode patterns for light end-fire coupled into such samples has clean intensity profiles in excellent agreement with calculated profiles. Losses, determined by a cutback technique for films on Si/SiO_xN_y substrates, were under 2 dB/cm with values as low as 1.6 dB/cm.

P4VP layers can be combined with polyvinylalcohol (PVA) and PMMA layers. P4VP/PVA alternating layer films were fabricated with up to eleven individual layers. The calculated and observed mode structures were in reasonable agreement, the observed differences being attributable to small variations in thickness of the individual layers. PMMA was used as a top layer in which ridge guides could be made by the addition of a depolymerising agent, benzil, and exposure to a high pressure mercury arc. Since the depolymerization products are volatile the process can be carried out in an air flow without any post exposure processing. Smooth ridges and ridge bifurcations were obtained by this method. Bifurcations (Y-junctions) are an essential structural component in a variety of integrated optic devices, e.g. Mach-Zehnder interferometers and multi-input and output switches, Nyman and Prucnal (1989). Other aspects accounts of this work have been reported in the literature, Wells and Bloor (1989a, b). ·

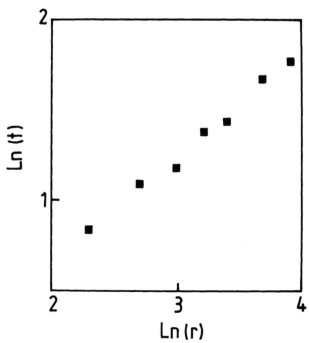

Figure 1. Dependence of film thickness (t) in microns on substrate withdrawal rate (r) in mm/min for P4VP films dip-coated from isopropylalcohol solution containing 165 g/l of polymer.

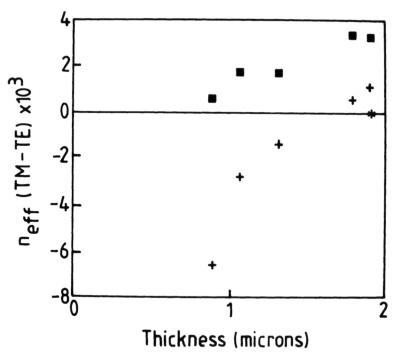

Figure 2. Mode index birefringence for P4VP dip-coated films as a function of film thickness: ■ (TM0-TE0), + (TM1 − TE1) and * (TM2 − TE2)

The production of multi-layer waveguide structures, ridge waveguides and clean end-faces without polishing, together with the use of a variety of substrates, illustrates the potential for the rapid and facile fabrication of complex waveguide structures utilising commercially available polymers. The potential this offers for the practical evaluation of optical waveguide configurations, which can be theoretically modelled but are difficult to prepare in conventional materials, has yet to be fully realised. This will, no doubt, be stimulated by the use of polymeric waveguides for the transfer of optical signals in hybrid devices, on printed circuit boards and in computer back-planes, Hartman et al., (1989). The use of planar, passive polymer structures in these areas is likely to increase in the near future as low loss polymers become commercially available. c.f Table 1. In this context the production of ridge waveguides by the use of uv and oxidative degradation of the Durham polyacetylene precursor polymer is of note, Gray (1989).

2.2 Thermal Sensors

The physical properties of polymers can be utilised to produce active optical waveguide devices. Specially developed polymers with large optical non-linearities are considered in Section 3. However, commercially available polymers can be used for other devices; notably sensors. Polymers in general have larger thermal expansion than inorganic materials and in consequence a larger temperature coefficient of the refractive index. Typical values of dn/dT for polymers, $1 - 4 \times 10^{-4}$ °C^{-1} are a factor of ten larger than those of inorganic materials. This difference has been utilised to produce thermally driven switches, Diemeer et al., (1989), and temperature sensing elements, Wells (1990) and Wells and Bloor (1989b). The latter are described in this section.

For a waveguide with little or no birefringence light can be coupled into TE and TM modes at the same synchronous angle. Equal intensity will be coupled into the TE and TM modes if the input light is polarized at 45° to the plane of incidence. An analyser in the output beam will give a maximum intensity for the same polarization and zero when rotated through 90°. This behaviour is readily demonstrated with P4VP films with zero birefringence, see Figure 2. Light was prism coupled into the TE0 and TM0 modes of such a film on a Si/SiO$_x$N$_y$ substrate and coupled out of a cleaved end-face after propagating 8.5 mm. On heating the film becomes birefringent. The difference in n(TE) and n(TM) probably results from anisotropic stress produced by the differential thermal expansion of the P4VP film and the substrate, i.e. there are contributions from dn/dT and stress induced optical anisotropy. Typical signal variation observed during the cooling of a preheated sample is shown in Figure 3.

A simpler and more sensitive structure was found for a 1.58 micron-thick P4VP film on a different Si/SiO$_x$N$_y$ substrate. The mode structure of this sample consisted of modes which were confined predominantly in the polymer and others with the bulk of the intensity in the silicon oxynitride buffer layer. The presence of the modes in the silicon oxynitride layer suggests that it has a non-uniform composition giving rise to a higher index region capable of supporting the modes. Though this was not achieved deliberately it could be reproduced by control of the chemical vapour deposition process used to make the buffer layer. End-fire coupling to a section cleaved from this sample showed approximately equal intensities in the polymer modes, A, B and C in Figure 4 and the buffer layer mode, D in Figure 4. Thus the sample can act as a Mach-Zehnder interferometer with one arm in the polymer the other in the buffer layer. On heating the sample the differential expansion of the P4VP and the silicon oxynitride produced a modulated output from a photodetector

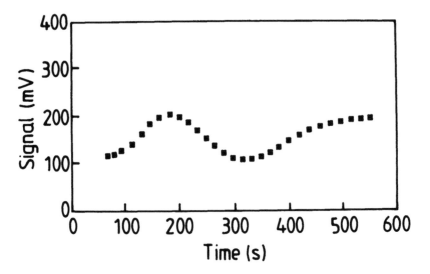

Figure 3.   Modulation of output due to thermally induced birefringence in a P4VP waveguide cooling under natural convection through 16.5°C.

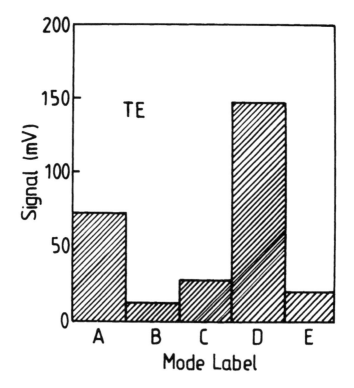

Figure 4.   Mode spectra of radiation end-fire coupled into a P4VP–SiO$_x$N$_y$ composite waveguide.   Mode index decreases from right to left.

at the focus of a lens collecting the light from the end of the guide. Figure
5 shows the temperatures at which intensity maxima were observed. Each
oscillation in intensity corresponds to a temperature change of about 1°C.
With precise intensity measurement and a longer path in the guiding structure
a temperature sensitivity of 0.01°C or less should be readily achieved.

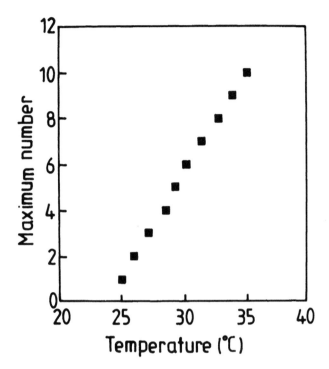

Figure 5. Experimentally determined temperatures for maxima of intensity for
light end-fire coupled into the P4VP-SiO$_x$N$_y$ composite waveguide with mode
spectra illustrated in Figure 4.

These two examples illustrate how by design of the optical waveguide the
sensitivity of the polymer layer to external variables can be used to provide
transducer action. In addition to temperature other variables which could be
sensed by such devices include humidity, organic vapours which swell the
polymer film, pressure, specific absorbates which change either the refractive
index or absorption of the polymer layer, etc. Butt coupling to optical fibre
of cleaved waveguide sections offers the prospect of easy assembly into
systems.

3. WAVEGUIDES IN RESEARCH POLYMERS

The potential areas of application for polymers possessing large optical non-
linearities include both in-plane propagation, for integrated optics, and
through plane propagation, for the control of light beams propagating in free
space, e.g. in optical computing, Ulrich (1989). The requirements for
materials for integrated optics are the more demanding. In addition to having
good non-linear optical properties the optical absorption and scattering of
the polymers must be very low to allow guiding over distances of a few
centimetres or more. A good long term stability is obviously also essential
for all applications. The attainment of these target properties is a

significant challenge, however, the potential rewards for success are high. Though the current generation of non-linear optical devices are based on lithium niobate, and other inorganic materials, polymers are anticipated to play an increasingly important role in a what, by early next century, will be a major market.

Research into polymers for non-linear optical devices has concentrated in two different classes of material. These are polymers with pendent groups with quadratic nlo properties for second harmonic generation and electro-optic devices, and those with conjugated backbones with cubic nlo properties for all optical applications, Stegeman et al., (1988) and Zyss (1985). These two classes are discussed separately in the following sections.

3.1 Quadratic nlo polymers

The prediction of the quadratic nlo properties of organic compounds using quantum chemical methods is now well established, Docherty et al., (1985) and Morley and Pugh (1989). However, the achievement of a macroscopic nonlinearity requires, in general, a non-centrosymmetric crystal structure. Since the requirement of low optical loss implies the use of amorphous polymers, which are isotropic, the quadratic non-linearity will be zero unless some polar order is induced in the polymer. This has been achieved principally by poling the polymer films with large electric fields. Electric field poling, either with evaporated electrodes or by a corona discharge, of PMMA films loaded with nlo active molecules has been extensively studied, Knabke et al., (1989), Kuzyk et al., (1989) and Mortazavi et al., (1989). Films poled near their glass transition temperatures display quadratic nlo properties when cooled to room temperature. Since the loading of nlo compounds that can be achieved is quite low the nlo coefficients obtained are modest. More damaging is the relaxation back to an isotropic material which occurs even well below the glass transition temperature.

To overcome this problem attention has switched to polymers in which the nlo moieties are attached as pendent groups. Such polymers can be obtained as either amorphous or liquid crystalline materials, Ulrich (1989). While the latter have been considered to offer better order and hence higher nlo coefficients, Mohlmann and van der Vorst (1989), in practise most groups have concentrated on the amorphous polymers, Matsumoto et al., (1987), Singer et al., (1988) and Dubois (1990). The behaviour of poled films of loaded polymers and copolymers with pendent active groups have been compared, Dubois (1990). The former give lower nlo coefficients which relaxed at a faster rate after poling. This is due to the lower fraction of the nlo active species and its greater freedom of motion in the loaded polymer film. The nlo coefficient of the copolymers depended linearly on the fraction of nlo active groups, and on the square of the poling field. Maximum nlo coefficient was achieved by poling just below the glass transition temperature of the polymer. Well below T_g there is insufficient molecular mobility for poling to be effective while above T_g the dielectric strength falls limiting the poling field. Initial non-linearities were high, in excess of these of lithium niobate, but were subject to a slow long-term relaxation. This was inhibited by the use of reactive pendent groups which allow the poling order to be locked in by post-poling photo-reaction.

Similar results have been reported from the USA, Ulrich (1989). However, for specific polymers thermal annealing can reduce the long-term relaxation to levels adequate for device applications. A number of different approaches to device design have been reported, Stegeman et al., (1987), Kaczmarski et al., (1989) and Cross et al., (1989). Kaczmarski et al., propose a novel structure in which confinement is achieved by a continuously applied field. Device structure is defined by metal top electrodes deposited on a buffer layer. This has the advantage of preventing relaxation in the active polymer layer. Electro-optical switching is achieved by applying a smaller differential potential to two closely spaced guides to affect the coupling between them. Unfortunately the practical implementation of these ideas was not discussed. In contrast Cross et al., 1989, consider the design and implementation of an electro-optic modulator using a pre-poled pendent group polymer. The confining channel was etched in a polymer buffer layer deposited on a gold on silicon substrate. Propagation and insertion loss as a function of channel width were determined. Amplitude modulation using a Mach-Zehnder geometry was demonstrated.

The rapid development of polymers with improved nlo properties and stability and progress in device design and implementation suggest that laboratory devices with competitive performance will be available in the very near future.

3.2 Cubic nlo polymers

Since there are no symmetry requirements for the occurrence of a finite cubic non-linearity a wider field of polymers is potentially available for investigation. The theoretical understanding of cubic non-linearity is less well established, Wu et al., (1989), Soos et al., (1990) and Kirtman and Hasan (1989). However, materials with polarizable electron systems have been identified as likely candidates for large cubic non-linearities. For polymers this means systems with conjugated backbones possessing extended π-electron systems. The polyenes and polyacetylene have, therefore, been taken as model systems for both theoretical and experimental study. In addition to polyacetylene, Drury (1988), cubic non-linearities have been determined for polypyrrole, Ghoshal (1989), polythiophene, Byrne et al., (1989), polyphenylacetylene, Neher et al., (1989); metal poly-ynes, Guha et al., (1989), poly-quinoxalines, Cao et al., (1989) and polydiacetylenes, Kanetake et al., (1989). The cubic non-linearities for most of these polymers fall in the range $10^{-11} - 10^{-12}$ esu. Exceptions are oriented Durham polyacetylene and single crystal polydiacetylene. The cubic non-linearities of these materials determined by third harmonic generation at $1.906\mu m$ are $2.7 \pm 0.4 \times 10^{-8}$ esu and $2.2 \pm 0.4 \times 10^{-9}$ esu respectively, Drury (1988).

Polydiacetylenes have been the subject of extensive fundamental and applied research. The latter has been reviewed, Etemad et al., (1990) and Bloor (1989), and only recent progress towards the realisation of devices, will be discussed here. Thin films suitable for optical waveguiding have been produced from the materials listed in Table 2. Use of a special growth

Table 2. Polydiacetylenes, general formula $\{C-C\equiv C-C\}_n$, used for optical waveguides

Abbreviation	R	Morphology
PTS	$-CH_2OSO_2(C_6H_4)CH_3$	Single crystal
nBCMU	$-(CH_2)_nOCONHCH_2COO(CH_2)_3CH_3$	Cast film
nSMBU/nRMBU	$-(CH_2)_nOCONHC^*H(CH_3)(C_6H_5)$	Cast film

technique has enabled single crystal films of PTS thin enough for use as optical waveguides to be fabricated. Channel waveguides have been photolithographically defined in these films. The electronic intensity dependent refractive index and intensity dependent phase shifts have been measured in these structures, Thakur and Krol (1990) and Krol and Thakur (1990). Spin coating of films of soluble polydiacetylenes such as 4BCMU produces glassy films that can act as optical waveguides, however, non-linear effects observed are predominantly thermal in origin, Etemad et al., (1990). Simple photobleaching techniques can be used to define channel guides in 4BCMU films, Rochford et al., (1989). The use of alignment layers to produce oriented films of 4BCMU capable of forming optical waveguides has been demonstrated, Patel et al., (1990).

The nRMBU and nSMBU polymers are chiral materials, Drake et al., (1989), which can be dip-coated to produce low loss slab and channel waveguides, Mann et al., (1988). Though intensity dependent effects have not been demonstrated in these guides they have been used to study the quadratic electro-optic effect, Oldroyd et al., (1989). These reveal that the non-linearity is reduced by the disorder in the polymer chains and the large fraction of the film occupied by the inactive pendent groups.

To date only the devices fabricated in the PTS single crystal films offer reasonable operating power and device size. Extrapolation of observed powers of about 1 watt for a π-phase shift indicates that devices switching at a few tens of milliwatt are possible. This is still rather high for high density switching devices, where power levels in the microwatt range are desirable. Thus, the realisation of practical polymeric all optical devices requires a considerable improvement in the cubic non-linearity.

4. Conclusions

Polymers are attractive materials for the fabrication of both passive and active integrated optical waveguides. Commercial materials are already available for passive optical interconnect and will play an increasing role in this area. Sensing functions can readily be achieved by choice of polymer and waveguide geometry. Such passive and active elements can be easily combined to address markets such as control systems for motor cars. Optically non-linear polymers and devices fabricated from them are still in the research laboratory. However, pendent group polymers for electro-optical applications are reaching a point where competitive devices are under development. Their commercial prospects depend on the realisation of good long term stability against thermally promoted depoling and photochemical degradation. In comparison conjugated polymers are still a long way from providing either the ease of fabrication or the magnitude of the cubic non-linearity necessary for the production of useful all optical devices. It should not be forgotten that materials that fail to meet the stringent requirements for optical waveguiding may well be useful in through plane applications.

Acknowedgements

Financial support from the Science and Engineering Research Council and GEC Marconi Research Centre is gratefully acknowledged.

References

Bloor, D., 1989, Molecular Electronics – Science and Technology, Ed. A. Aviram (Eng. Found, New York) p.259.
Brown, B. L., 1989, J. Lightwave Techn. 7, 1445.

Byrne, H. J., Blau, W. and Jen, K-Y, 1989, Synth. Metals 32, 229.

Burzynshi, R., Singh, B. P., Prasad, P. N., Zanoni, R. and Stegemen, G. I., 1988, Appl. Phys. Lett. 53, 2011.

Cao, X. F., Jiang, J. P., Bloch, D. P., Hellworth, R. W., Yu, L. P. and Dalton, L, 1989, J. Appl. Phys. 65, 5012.

Chemla, D. S. and Zyss, J., 1987, Nonlinear Optical Properties of Organic Molecules and Crystals (Academic Press, Orlando).

Cross, G. H., Donaldson, A., Gymer, R. W., Mann, S., Parsons, N. J., Haas, D. R., Man, H. T. and Yoon, H. N., 1989, Proc. SPIE 1177, 79.

Diemeer, M. B. J., Brons, J. J. and Trommel, E. S. 1989, J. Lightwave Techn. 7, 449.

Docherty, V. J., Pugh, D. and Morley, J. O., 1985., J. Chem. Soc. Faraday, Trans. 2. 81, 1179.

Drake, A. F., Udvarhelyi, P., Ando, D. J., Bloor, D., Obhi, J. S. and Mann, S., 1989, Polymer 30, 1063.

Driemeier, W., 1990, Opt. Commun. 76, 25.

Driemeier, W. and Brockmeyer, A., 1986, Appl. Opt. 25, 2960.

Dubois, J. C., 1990, Conjugated Polymeric Materials, Ed. Bredas, J. L. and Chance, R. R. (Kluwer Academic Publ, Dordrecht) p.321.

Drury, M. R., 1988, Sol. St. Commun. 68, 417.

Etemad, S., Fann, W-F, Townsend, P. D., Baker, G. L. and Jackel, J., 1990, Conjugated Polymeric Materials, Ed. Bredas, J. L. and Chance, R. R. (Kluwer Academic Publ., Dordrecht) p.341.

Franke, H., Knabke, G. and Reuter, R., 1986, Proc. SPIE 682, 191.

Ghoshal, S. K., 1989, Chem. Phys. Lett. 158, 65.

Gray, S. 1989, Photonics Spectra Sept. p.125.

Guha, S., Frazier, C. C., Porter, P. L., Kang, K. and Finberg, S. E., 1989, Opt. Lett. 14, 952.

Hann, R. A. and Bloor, D., 1989, Organic Materials for Non-linear Optics (Royal Soc. Chem., London).

Hartman, D. H., Lalk, G. R., Howse, J. W. and Krchnavek, R. R., 1989, Appl. Opt. 28, 40.

Heeger, A. J., Orenstein, J. and Ulrich, D. R. 1988, Nonlinear Optical Properties of Polymers (Mat. Res. Soc., Pittsburgh).

Kaczmarski, P., van de Capelle, J-P, Lagasse, R. E. and Meynart, 1989, IEE Proc. Pt. J. 136, 152.

Kaino, T. 1986, Appl. Phys. Lett. 48 757.

Kanetake, T., Ishikawa, K, Hasegawa, T., Koda, T., Takeda, K., Hasegawa, M, Kubodera, K. and Kobayashi, H., 1989, Appl. Phys. Lett. 54, 2287.

Kapoor, S. K., Pandey, C. D., Joshi, J. C., Dawar, A. L. Tripathy, K. N. and Gupta, V. L., 1989, Appl. Opt. 28, 37.

Kirtman, B. and Hasan, M., 1989, Chem. Phys. Lett. 157, 123.

Knabke, G., Franke, H. and Frank, W. F. X., 1989, J. Opt. Soc. Amer. B. 6, 761.

Krol, D. M. and Thakur, M., 1990, Appl. Phys. Lett. 56, 1406.

Kuzyk, M. G., Singer, K. D., Zahn, H. E. and King, L. A. 1989, J. Opt. Soc. B. 6, 742.

Mann, S. Oldroyd, A. R., Bloor, D., Ando, D. J. and Wells, P. J., 1988, Proc. SPIE 971, 245.

Matsumoto, S., Kubodera, K., Kurihara, T. and Kaino, T. 1987, Appl. Phys. Lett. 51, 1.

Mohlmann, G. R. and van der Vorst, C. P. J. M., 1989, Side Chain Liquid Crystal Polymers, Ed. McArdle, C. B. (Blackie, Glasgow), p.330.

Morley, J. O. and Pugh, D., 1989, Organic Materials for Non-linear Optics, Eds. Hann, R. A. and Bloor, D., (Royal Soc. Chem., London) p.28.

Mortazavi, M. A., Knoesen, A., Kowel, S. T., Higgins, B. G. and Dienes, A., 1989, J. Opt. Soc. Amer. B.6, 733.

Neher, D., Wolf, A., Bubeck, C. and Wegner. G. 1989, Chem. Phys. Lett. **163**, 116.

Nyman, B. M. and Prucnal, P. R., 1989, Opt. Eng. **28**, 982.

Okamoto, N. and Tashiro, S. 1988, Opt. Commun. **66**, 93.

Oldroyd, A. R., Mann, S. and McCallion, K. J., 1989, Electr. Lett. **25**, 1476.

Patel, J. S., Lee, S-D., Baker, G. L. and Shelburne, J. A. III, 1990, Appl. Phys. Lett. **56**, 131.

Prasad, P. N. and Ulrich, D. R., 1988, Nonlinear Optical and Electroactive Polymers (Plenum Press, New York).

Prest, W. M. Jnr. and Luca, D. J., 1979, J. Appl. Phys. **50**, 6067.

Reuter, R., Franke, H. and Feger, C., 1988, Appl. Optics. **27**, 4565.

Rochford, K. B., Zanoni, R., Gong, Q. and Stegeman, G. I., 1989, Appl. Phys. Lett. **55**, 1161.

Savatinova, I., Tonchev, S., Todorov, R., Venkova, E., Liarokapis, E. and Anastassakis, E., 1990, J. Appl. Phys. **67**, 2051.

Schriever, R., Franke, H., Festl, H. G. and Kratzig, E. 1985, Polymer **26**, 1423.

Singer, K. D., Kuzyk, M. G., Holland, W. R., Sohn, J. E., Lalama, S. J., Comizzoli, R. B., Katz, H. E. and Schilling, M. L., 1988, Appl. Phys. Lett. **53**, 1800.

Soos, Z. G., Hayden, G. W. and McWilliams, P. C. M., 1990, Conjugated Polymeric Materials, Ed. Bredas, J. L. and Chance, R. R. (Kluwer Academic Publ., Dordrecht) p. 495.

Stegeman, G. I., Seaton, C. T. and Zanoni, R., 1987, Thin Solid Films **152**, 231.

Thakur, M. and Krol, D. M., 1990, Appl. Phys. Lett. **56**, 1406.

Ulrich, D., 1989, Organic Materials for Non-linear Optics, Eds. Hann, R. A. and Bloor, D. (Royal Soc. Chem, London) p.241.

Wells, P. J., 1990, Thesis Univ. of London.

Wells, P. J. and Bloor, D., 1989a, Organic Materials for Non-Linear Optics, Eds. Hann, R. A. and Bloor, D. (Royal Soc. Chem, London) p.398.

Wells, P. J. and Bloor, D., 1989b, Proc. SPIE **1018**, 58.

Wu, J. W., Heflin, J. R., Norwood, R. A., Wong, K. X., Zamani-Khamiri, O., Garito, A. F., Kalyanaraman, P. and Sounik, J. 1989, J. Opt. Soc. Amer. B. **6**, 707.

Yang, C-C, Josefowicz, J. Y. and Alexandru, L., 1980, Thin Solid Films **74**, 117.

Zyss, J., 1985, J. Molec. Electron. **1**, 25.

Conjugated polymers—prospects for semiconductor device physics

R. H. Friend, J. H. Burroughes and K. E. Ziemelis

The Cavendish Laboratory, Madingley Road, Cambridge CB3 0HE, UK

ABSTRACT: Conjugated polymers which can be processed to form fully dense thin films can be used as the active component in conventional semiconductor device structures such as the MISFET. Devices of this type can show low densities of defects states within the gap, and we discuss how this may be the case for these polymeric materials which clearly show considerable disorder At a microscopic level it is necessary to consider the role of self localisation of carriers injected onto the chains to form polarons (or for the case of *trans*-polyacetylene, solitons). This self-trapping of charge onto a specific chain is disadvantageous for charge transport; the rate limiting step for electronic conduction is set by interchain charge transfer which is a thermally activated process, and mobilities are typically very low, of order 10^{-4} $cm^2V^{-1}s^{-1}$. However, within these self-trapped excitations there is considerable reorganisation of the π-electron structure, and states are pushed from the conduction and valence bands into the semiconductor gap, giving rise to large changes in the sub-gap optical absorption.. This gives novel electro-optical behaviour for devices fabricated with conjugated polymers, with control of the sub-gap absorption through for example the charge density within the accumulation layer in an MIS or MISFET structure.

1. INTRODUCTION

Although conjugated polymers have been available in a form suitable for many physical measurements for some time, it is only comparatively recently that we have had access to polymers which can be processed to form thin, uniform, dense and coherent films, which is the basic requirement for attempting to use these materials as the active elements in semiconductor devices. The principal reason for this is that most conjugated polymers cannot be directly processed to the forms required in these devices, being not readily soluble in easily-handled solvents and infusible. This has severely limited the scope for construction of devices (Grant et al 1980, Kanicki 1986). Improvements in the processing of conjugated polymers, through electrochemical deposition during polymerisation, and the use of solution-processible poly(3-alkyl thiophenes) has allowed better device fabrication (Garnier and Horovitz 1987, Tomozawa et al 1987), and the use of model oligomers which can be deposited by vacuum sublimation has recently been reported (Garnier et al 1989). There are also reports of field-induced conductivity measured in MISFET (Metal-Insulator-Semiconductor Field Effect Transistor) structures (Ebisawa et al 1983, Koezuka et al 1987,1989, Assadi et al 1988).

A major advance in the control of the polymer processing is in the use of a solution-processible precursor polymer which can be converted to the conjugated polymer after processing. This was first demonstated by Edwards and Feast (1980), for the preparation of polyacetylene (the Durham route), see also Edwards et al (1984) and Feast and Winter (1985), and it is polyacetylene produced by this route that we have used for much of our work. We make thin-film devices by spin-coating the precursor polymer, poly((5,6-bis(trifluoro-methyl)-

film devices by spin-coating the precursor polymer, poly((5,6-bis(trifluoro-methyl)-bicyclo[2,2,2]octa-5,7-diene-2,3-diyl)-1,2-ethene diyl), in solution in 2-butanone onto the required substrate, followed by heat treatment at typically 100°C to convert to the polyacetylene, by elimination of hexafluoroorthoxylene. The precursor polymer has very good film-forming properties, and it is straightforward to control the thickness of the resultant polyacetylene film over the range 100Å to several μm. We have investigated a range of unipolar devices: Schottky-barrier diodes, MIS (Metal-Insulator-Semiconductor) structures and MISFET's (Burroughes et al 1888,1989a,1989b,1990, Lawrence et al 1989). By careful control of the processing, with rigorous exclusion of oxygen, we have been able to get the device performance up to levels respectable enough to learn about the detailed functioning of the device. More recently we have been working with some soluble poly(3-alkyl thienylenes) on which we have recently carried out a study of the photoexcitation spectroscopy (Rühe et al 1990).

Interest in the semiconductor device physics of conjugated polymers comes both from the possibility for finding new types of structure which might be of technological use, and also in the use of these structures for the investigation of the basic physical processes that determine the device characteristics. In effect, suitable devices provide a new way of introducing excitations into these polymeric semiconductors. Chemical doping and photoexcitation of electron-hole pairs are both well studied routes for the conjugated polymers, but the construction of semiconductor device structures such as the MIS (Metal Insulator Semiconductor) device, offers a third method, through the injection of charge to form regions of space charge density (as in the depletion layer) or surface charge density (as in accumulation or inversion layers in the MIS device).

It is well known that the response of the coupled electron-lattice system for conjugated polymers is non-linear, and that for example charged excitations of the chain take the form of self-localised polarons (or for the case of *trans*-polyacetylene, as solitons) Su et al (1979,1980, Fesser et al 1983). The formation of these states modifies completely the model we must use for the semiconductor physics; we identify two aspects of this here. Firstly, charge transport requires that these excitations move, and we must distinguish between intra-chain motion which may still be band-like and inter-chain motion which as we discuss later in the paper requires thermal activation to transfer the excitation from one chain to another. This latter process is in general likely to limit the performance we can expect from these devices. Secondly, these self-localised excitations are known to pull energy states from the π bands into the region of the band gap, so that new optical transitions between these gap states and the band edges are possible. Transition matrix elements for these transitions are large, and we can therefore expect to see relatively large electro-optical responses in these polymers, in regions of photon energy well away from the interband absorption and in which the optical losses through the polymer in the absence of the injected charges are very low. For the case of polyacetylene, the excited states produce non-bonding π^* states, which lie at mid-gap, to accommodate the injected charge. Thus we expect that charge storage in the polymer will be in soliton states created in sufficient number to store all the injected charge. We expect to see the Fermi energy pinned close to the centre of the gap, and to see the density of soliton states vary with external field. Besides the range of electrical measurements that characterise the device performance, we have carried out an extensive range of optical measurements, since the clearest evidence for the formation of solitons in polyacetylene is through the appearance of additional optical absorption below the band gap, associated with the vibrational and electronic excitations of the solitons.

2.   CONDITIONS FOR DEVICE OPERATION

Devices such as the Schottky barrier diode and the MIS and MISFET structures depend for their operation on the control of charge density within the semiconductor through the external gate voltage. In the MIS and MISFET devices, the gate is separated from the active semiconductor by a thin insulator layer with a high dielectric strength, and the applied gate voltage is used to create a very high electric field at the interface between the semiconductor and insulator. This can be used to bend energy bands in the semiconductor towards or away from the Fermi energy. We follow the example of the p-type semiconductor applicable to the polyacetylene as used in our work. With a large negative gate voltage, the valence band is pulled upwards, and if it crosses the Fermi energy, holes are created in the surface layer of the semiconductor. This surface charge layer is known as the accumulation layer. As the gate voltage is reduced and the accumulation layer is removed, a depletion layer is formed in which the mobile charges associated with the extrinsic dopants (acceptors) are pushed away from the interface. If the gate voltage is further raised, then it may be possible to lower the conduction band towards the Fermi level, and allow occupation of conduction band states at the interface, to form a charge inversion layer of n-type carriers. The MIS structure is particularly versatile, and for our purposes has the special advantage that in the accumulation and inversion modes, the charge injected is present without the countercharge from a dopant species, and we can hope to make measurements on carriers which are not perturbed by the presence of dopants between the chains.

There are stringent conditions that must be met before a semiconductor can be used in this type of device. Firstly, the semiconductor must be free of defects that have associated energy levels within the semiconductor gap. Polymers do of course possess many defects from regular crystalline order, and for the polyacetylene that we use, prepared by the Durham precursor route, we have many chain bends or twists that interrupt the straight chain sequences. However, we can expect quite generally that the degree of conjugation at a bend or twist will be lowered, and we can expect that the associated π and π^* states will move deep into the bands.

Secondly, the interface between the insulator and the semiconductor must be free of surface states, caused for example by dangling bonds at the semiconductor surface. The maximum surface charge density that can be achieved in the accumulation or inversion layer in the MIS device is fixed by the dielectric strength, E_b of the insulator, as $n_{max} = \varepsilon_0\varepsilon_rE_b/q$ where ε_r is the relative dielectric constant. With silicon dioxide as the insulator, this gives of order 2×10^{13} electrons/cm^2; this is a low value in comparison with the 10^{15} or so carbon atoms/cm^2 at the polyacetylene surface. Polymers do not, of course, have unsatisfied bonds at the surface, since bonds are only formed along the polymer chain, and we can expect that if the structure can be fabricated without inclusion of contaminants in the interfacial region, the device should be free of surface states. A likely source of contamination is from included oxygen, and it is probably always necessary to process polymeric semiconductor devices in the absence of oxygen.

Thirdly, fabrication of semiconductor devices requires control of the concentration of the extrinsic charge carriers, typically in the concentration range 10^{15} to 10^{18} charges/cm^2 (this is in order that widths of depletion regimes are of the correct size, of order 0.1 to 1 µm) with the further requirement that the dopant species be immobile at the operating temperature of the device. This poses a major challenge for the use of conjugated polymers in these applications because the chemical and electrochemical methods commonly used to 'dope' conjugated polymers, in which doping is achieved by diffusion of dopants through the solid polymer at room temperature, are quite inappropriate here. Apart from the problem of dopant mobility, it is very difficult to control doping levels at the very low dopant concentrations required (10^{-6} molar %), and at such low levels it is very unlikely that the doping would be homogeneous.

In the case of the Durham route polyacetylene that we use in this work, the polymer as made is extrinsically doped with p-type carriers in a convenient concentration so that we are able to use it without further treatment and we find that these dopants do not move under the applied field. We consider that they may be associated with catalyst residues at the chain ends of the precursor polymer, but the concentrations are too low to establish this by chemical means. For other polymers we have no such simple method to give the control we need, and in general it will be necessary to develop new methods for doping. For the poly(3-alkyl thienylenes) which are prepared in the oxidised state by chemical or electrochemical oxidation of the monomer, and which are subsequently reduced to the 'undoped' state, it has been found that there is a residual versus bias voltage in Schottky or MIS diodes is typically in the range 10^{17} cm^{-3}. We find however that there is clear evidence that the dopants responsible for these carriers are mobile, and that this motion does interfer with the operation of the device.

3.   DEVICE FABRICATION AND ELECTRICAL CHARACTERISATION

For 2-terminal devices, such as the Schottky diode and the MIS structure, the device is fabricated as a series of layers. For example, to construct an MIS structure we have used silicon as substrate with a doped layer to act as gate and a native oxide layer on top as the insulator. This structure is then coated with polymer and finally capped with an evaporated thin gold top contact. For the MISFET structure it is convenient to put the source and drain contacts onto the insulator layer, and put the polyacetylene layer down on top as the last stage in the construction. In the example of the MISFET structure is shown in figure 1, the source and drain contacts are of poly n-silicon; we have also made structures such as these with gold.

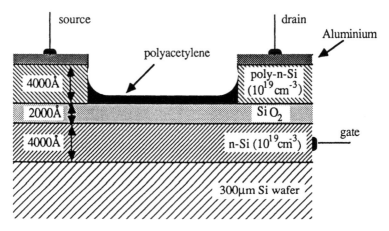

Figure 1   Schematic diagram for a polyacetylene MISFET structure. Dimensions shown are to scale, except the channel width (20 µm) and length (1.5 m).

The MIS structure allows the possibility of band bending at the insulator-semiconductor interface through the Fermi level to produce a surface charge layer which may be the same carrier sign as the majority carriers (accumulation layer) or as the minority carriers (inversion layer) (Sze 1981). The formation of accumulation, depletion and inversion layers may be demonstrated through the behaviour of the device capacitance with respect to the bias voltage. The measured capacitance, C, is that of the series combination of the insulator capacitance, C_i, and the capacitance of the active region of the semiconductor, C_d, and is given by $C = C_iC_d/(C_i+C_d)$. Since C_d is large for the accumulation and inversion layers, C is equal to the geometric capacitance of the insulating layer, but C falls to a lower value for movement of the depletion layer. The capacitance versus bias voltage curve for a polyacetylene MIS structure is shown in figure 2. As expected, the capacitance for negative gate voltages flattens out to the geometric capacitance of the insulator, indicating the formation of an accumulation layer. The decrease in the measured capacitance for positive voltages displays the formation of the depletion layer. Saturation of the capacitance for large positive biases sets in when the depletion layer extends across the polyacetylene film (estimated to be about 1200Å for this device).

We show the capacitance versus bias voltage for a similar structure fabricated with poly(3-hexyl thienylene) in place of polyacetylene in figure 3. We note that the structure behaves in a similar way to the polyacetylene MIS device, with a clear transition from depletion at positive bias to accumulation for negative bias. However, the voltage at which the devices switches from one mode to the other is dependent on the previous history of the device, as seen by the large hysteresis in voltage for the transition for rising and falling bias voltage. The time scale for the movement of the threshold voltage is of order minutes and we consider that this is due to the migration of ionic charge in the applied electric field. For the poly(3-hexyl thienylene) we used here, which was prepared by chemical oxidation of 3-hexyl thiophene, using ferric chloride, we presume that the ions present and able to migrate are $FeCl_4^-$. Similar evidence for ionic motion in this polymer has been obtained for Schottky diode structures by Inganäs (1990).

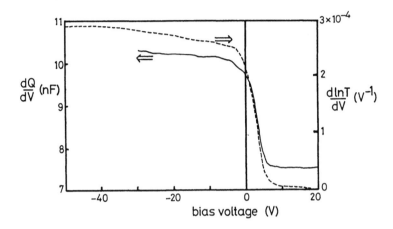

Figure 2 The differential optical transmission, $\partial lnT/\partial V$ at 0.8 eV, and differential capacitance, $\partial Q/\partial V$ versus bias voltage for a polyacetylene MIS structure. Measurements were made at 500 Hz and with an a.c. modulation of 0.25 eV.

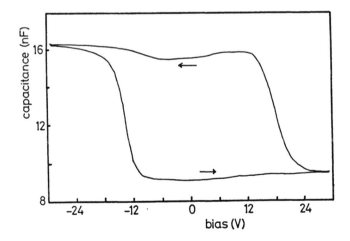

Figure 3 Differential capacitance, $\partial Q/\partial V$ versus bias voltage for a poly(3-hexyl thienylene) MIS structure. Measurements were made at 250 Hz and with an a.c. modulation of 0.25 eV. Sweeps of the bias voltage were carried out at about 6 V/minute.

Demonstration that the charges at the polyacetylene/insulator interface in the MIS structure are mobile is made in a MISFET structure, in which source and drain contacts allow measurement of the conductance of the surface charge layer. The choice of the material used for construction of the source and drain contacts should determine whether the device operates in inversion or accumulation mode. The MISFET shown in figure 1 is constructed with n-silicon source and drain contacts, which might be expected to make ohmic contacts to n-type polyacetylene, and thus give enhanced channel conductance from formation of an n-type inversion layer for positive gate voltages. Figure 4 shows the variation of I_{DS} with V_{GS} at constant V_{DS}. The channel conductance has a minimum at $V_{GS} = +10V$ (full depletion) and rises for gate voltages both more positive (inversion layer) and more negative (accumulation layer), with a maximum on/off ratio for this structure of 100,000 ($V_{GS} = -40V$ to $V_{GS} = +10V$). We see, therefore, the behaviour expected for positive bias, though the strong enhancement of the conductivity for negative bias indicates that the accumulation layer is still able to make good contact to the source and drain electrodes. It is unusual to find a device able to operate in both modes, and we consider that the n-silicon work function matches that of the mid-gap 'soliton' band in polyacetylene, and that in contrast to traditional semiconductors, the Fermi level does not move far from the mid-gap band with addition of either p or n carriers.

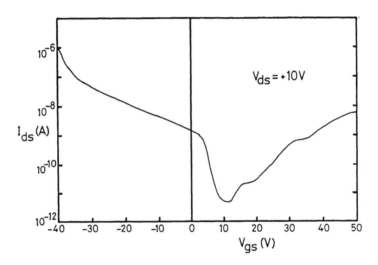

Figure 4. I_{DS} versus V_{GS} at constant V_{DS} (+10V) for a polyacetylene MISFET with poly n-silicon source and drain contacts, as shown in figure 1.

The MISFET structure provides a very convenient means of controlling the charge carrier concentration in the surface layer of the semiconductor. We have determined the carrier mobility from conductance measurements as a function of gate voltage (Burroughes et al 1989b), and find low values, typically 10^{-4} cm^{2}V^{-1}s^{-1}. Similar values for the mobility are estimated for photogenerated carriers, and the extrinsic carriers in the as-made polyacetylene (Townsend and Friend 1989). We discuss mechanisms for charge transport further in §5.

4   SELF-LOCALISATION OF INJECTED CHARGE

As discussed in §1, charge is stored in polyacetylene in 'soliton' localised states, which have non-bonding p_z 'mid-gap' states associated with them. We expect to see, therefore, a modulation in the optical properties of the active semiconductor region of these devices with applied bias voltage. The devices described above have all been constructed so that it is possible to pass sub band-gap light through the structure and thus to observe the changes in optical properties as the gate voltage is varied. For the MIS and MISFET structures we expect a decrease in the device transmission below the band-edge as the device is driven towards accumulation, and new charged solitons are introduced onto the polyacetylene chains at the interface with the insulator. The experiment is easily performed by modulating the gate voltage at a convenient frequency and monitoring with a lock-in amplifier the optical transmission signal that is modulated by the gate voltage. The electromodulation spectrum between 0.4 and 1.2 eV in figure 5 shows that there is a decrease in the transmission through the device for negative bias. The peak value of $\Delta T/T$ is at 0.8 eV. If all the charge at the polymer-insulator interface is stored in soliton-like states then $\partial \ln T/\partial V$ should scale with $\partial Q/\partial V$, the differential capacitance, as the bias voltage is varied. This is shown to be the case in figure 2, in which $\partial \ln T/\partial V$ and the differential capacitance are both plotted against bias voltage. The ratio of $\partial \ln T/\partial V$ and $\partial Q/\partial V$ gives the optical cross-section per charge injected, and in the accumulation regime we find a value at the peak in the 'mid-gap' absorption as 1.2×10^{-15} cm^2, in very close agreement with the value measured from dopant-induced absorption (Orenstein 1986).

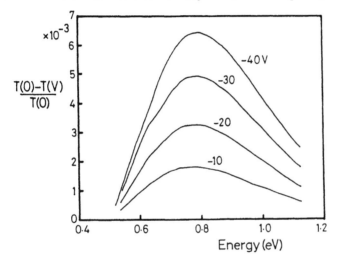

Figure 5. Voltage-modulated transmission for a polyacetylene MIS diode, [T(0) - T(V)] / T(0) versus the photon energy, for various values of V (all negative with respect to the gate).

Besides the electronic transitions associated with the soliton state, there is extra IR activity due to new vibrational modes which couple to motion of the charged soliton along the polymer chain. These are the translation modes of the soliton, and are seen in the polyacetylene accumulation layer at 1379 and 1281 cm^{-1} (the lowest mode is obscured by absorption by the silicon substrate) (Burroughes et al 1988, 1990). There are additional vibrational modes of the soliton, including the width-modulating amplitude mode which is Raman-active. This again has been observed in the polyacetylene accumulation layer (Lawrence et al 1989) at a frequency about 50 cm^{-1} lower than the corresponding modes of the dimerised chain.

5. CHARGE TRANSPORT PROCESSES

Charge transport in doped conjugated polymers has been extensively investigated but remains poorly understood. There was considerable early interest in the nature of the 'insulator-metal' transition achieved by chemical doping. It was observed that the conductivity reaches a metallic level at about 1% doping whereas the magnetic susceptibility remains low up to about 6% doping, at which level a transition to a Pauli-like behaviour is seen. There was considerable discussion as to whether the conduction processes in this intermediate regime were due to sliding of self-localised charged solitons along the polymer chain, but it is now clear that the transitions seen at around 6% doping are due at least in part to a phase transition between different orderings of the dopant ions around the chains (Heeger 1986).

In the limit of low charge carrier concentrations it is clear that carrier mobilities are low, and it is generally accepted that the rate-limiting process is that of charge transfer between chains. For the particular case of *trans*-polyacetylene, the topological character of the soliton raises some problems here, since it cannot move from one chain to another. There are two models developed for interchain charge transfer. The first of these is due to Kivelson (1982) and relies on the presence of a substantial concentration of neutral 'soliton' states. The process of interchain charge transfer is then achieved by transfer of charge from a charged soliton on one chain to a neutral soliton on an adjacent chain. This process, termed 'intersoliton hopping' is clearly specific to polyacetylene, both because of the particular nature of the soliton excitation and because of the requirement for both charged and neutral solitons. The second model is one in which the charged solitons are present in pairs on a single chain, and constrained (by for example interchain coupling or finite chain lengths) to remain close to one another. The doubly charged excitation may then be treated as a bipolaron, and the process of interchain charge transport is then controlled by the rate at which these can hop (or possibly tunnel) between chains (Townsend and Friend 1989). We consider that there is very strong evidence that this is the correct mechanism, for charge transport for carriers introduced through chemical doping, photoexcitation, and through charge injection.

The intersoliton model depends crucially on the presence of 'neutral solitons' which can be readily ionised and thus participate in the charge transport. Trans-polyacetylene, produced either by the Shirakawa route or by the Durham route, does contain a high concentration of spin defects which are detected in ESR experiments, and which can be shown to be due to the presence of non-bonding p_z states with wavefunctions spread over some 10 to 20 carbon sites. These spins appear during the thermal isomerisation from cis to trans, and are present typically in concentrations of 10^{18} - 10^{19} cm^{-3}. However, the bulk of the evidence gathered from photoexcitation experiments is that though they may be 'soliton-like' in the nature of their wavefunctions, they do not possess energy levels within the gap, and are not involved in the production of metastable photoexcited states (Townsend and Friend 1987,1989, Colaneri et al 1988). Turning to the experiments we have performed with charge accumulation layers, we are able to achieve rather higher levels of charge injection than obtained in the photoexcitation experiments. At the higher gate voltages used in the data shown in figure 5, the surface charge density is up to 4×10^{12} cm^{-2} (at -40 V). We consider that the charge accumulation layer is strongly localised to the interface with the insulator, probably within 2 nm. The total areal concentration of the neutral spin defects within such a layer is a factor of 2 lower than the charge concentration (at V_g = -40 V), and we do not therefore consider that the charge injected into the polymer is stored on the spin defects. As previously discussed by Townsend and Friend (1989) and Colaneri et al (1988), we consider that the neutral spin defects are present on disordered regions of the chain, which possess a high π-π^* energy gap, and thus excluded from participation in the electronic processes that take place within the gap.

The 'bipolaron model' is generally applicable for conjugated polymers, whether or not they may in principle possess a degenerate ground state. As discussed by Chance et al (1984), like-charged soliton pairs are analogous to bipolarons and are not topologically restrained from inter-chain hopping. The process by which they do this will involve an intermediate stage in which one of the two charges has tranferred to an adjacent chain, and the instantaneous description is of two polarons on adjacent chains. If the second charge then follows the first, the bipolaron has moved from one chain to another, and has surmounted an energy barrier equal to the stabilization energy of the bipolaron (or soliton anti-soliton pair). Townsend and Friend (1989) show that there is a well-defined activation energy for the mobility of photocarriers, of about 0.31 eV, and identify this as the bipolaron stabilisation energy. The general expression for the hop rate of a self-localized carrier is $R = \omega \exp\{-W_h/k_BT\}$ where W_h is the hop energy and ω is the attempt-to-hop frequency (Mott 1987), and this is shown to be consistent with the room temperature mobilities observed if the average distance between hops is set at 2.1 nm.

For the case of photocarriers, Townsend and Friend were able to identify from spectroscopic measurements that they are present as like-charged pairs of solitons; hence the description of the transport in terms of bipolaron motion along and between chains. For carriers introduced through chemical doping the situation is less clear because in general we have less spectroscopic information. It is interesting to compare the properties of as-made samples of Durham polyacetylene. These contain the usual level of spin 1/2 defects, as already discussed, and they also contain a low concentration of charges which are responsible for the DC conductivity. Measurements on Schottky barrier diodes formed between aluminium and the polyacetylene (Burroughes et al 1988) do provide the necessary information to characterise the charges present. From the current/voltage and capacitance/voltage characteristics we know that the charge carriers are p-type and present in a concentration of about 10^{16} cm^{-3}, and that this does not vary strongly with temperature. From the differential optical transmission measurements through the depletion regime we also know that these p-type carriers show the same 'mid-gap' optical absorption signature that are found in photo-induced absorption for the photocarriers, with a peak in absorption at 0.55 eV. The mobility of the dark carriers is now easily fixed by the value of the dark conductivity. The room temperature carrier mobility inferred from the conductivity (3 x 10^{-8} S/cm) is about 2 x 10^{-5} cm^2/Vs, and it shows an activation energy of about 0.4 eV (Townsend and Friend 1989). These values are rather similar to those we have stablished for the photocarriers, and we consider that the same model for transport is appropriate for both the extrinsic carriers and the photocarriers.

Turning to the carriers present in the accumulation and inversion layers in the MISFET structures, we have established that the mobilities are comparable to those of the extrinsic carriers in the as-made polyacetylene and to those of the photocarriers. The MISFET structure provides a very versatile system for the control of the carrier density and for the systematic measurement of the transport properties, and this is an area for future study. It is important to recognise that the accumulation or inversion layer in the MISFET structure is confined to the interface between the polymer and insulator, and is thus very sensitive to the surface structure of the polymer. The optical characterisation of the self-localised states created to store the injected charges that we are able to perform gives a very powerful method for characterising the π electron system of the polymer chains on which the charge is stored. We have previously noted (Burroughes et al 1990) that the 'mid-gap' electronic absorption in these structures is very dependent on the construction of the device. For devices which use silicon dioxide as the insulator layer the 'mid-gap' electronic absorption is at 0.8 eV. This is high, well above the value of 0.55 eV found for photo-excited charges in the bulk (Townsend and Friend 1989), and indicates that the local band gap for the polyacetylene chains which form the surface layer is higher than in the bulk. In contrast, devices built with a polymer, such as poly(methylmethacrylate) or poly(imide) as the insulator show a 'mid-gap' electronic absorption at 0.55 eV, and we consider that the polyacetylene formed at this interface is no more disordered than in the bulk (Burroughes and Friend 1990). Clearly, control of the electronic structure of the polyacetylene on which the accumulation layer is formed is important in the study of electronic transport processes in these surface charge layers. We are currently extending our measurements to tackle this important area.

6. PROSPECTS FOR FUTURE DEVELOPMENT

The work that we have discussed above gives a firm base for future development of polymeric semiconductor devices; there is clear demonstration that devices can be built without any great difficulties and at low cost, and which operate very much as we would expect. It is very encouraging to see for example that the field effect in the MIS structure can be obtained without problems due to defect states in the gap, either in the bulk or associated with the interface at the semiconductor/insulator interface. However, there is much still to be done before devices of this type will find their place in the market. We comment here on two aspects of our work that are relevant here.

6.1   Carrier Mobilities

We have in §5 discussed at some length the modelling of the charge transport processes that we consider to be appropriate for conduction in the regime of carrier concentration encountered in devices such as the MISFET, and we can draw some general conclusions for transport that requires thermal activation to move charge from one localised state to another (in this context, the inter-chain hop). For three-dimensional hopping, the mobility obtained using the Einstein diffusion relation is

$$\mu = \frac{ea^2\omega}{6k_BT} \exp\{-W_h/k_BT\}$$

where a is the hop distance and W_h is the barrier to hopping as discussed in §5 above (Mott 1987). If we take a = 0.42 nm (the interchain separation for polyacetylene), $\omega = 3.9 \times 10^{13}$ Hz (the average frequency of the optic phonons which modulate the chain dimerization) and $W_h = 0.31$ eV (the measured activation energy), then at 290 K we obtain a mobility of 1.9×10^{-6} cm^2/Vs. This value is lower than that measured and reason for this is that the hop distance parameter, a, is more correctly interpreted as the average distance that a carrier is able to travel between interchain hops. For the case of conjugated polymers, this distance is not the interchain separation, but the average distance that carriers are able to travel <u>along the chain</u> between the thermally-activated inter-chain transfer processes. This distance may be associated (loosely) with the so-called chain conjugation length. In order to fit the measured value for the unoriented polyacetylene used here a value of a = 2.1 nm is required and we consider that this is a reasonable value for unoriented Durham polyacetylene (Brown et al 1986).

We can ask what the prospects are for increasing mobilities to values that are comparable to those found for amorphous silicon FETs. There are two points to consider. Firstly, if transport is controlled by the equation for mobility given above, the only parameter that is experimentally adjustable is the distance between interchain hops, a. We can expect to be able to improve on this value very considerably if we can find ways to improve the structural order in the active region of the polymer, i.e. at the interface with the insulator layer in the MIS structure. There is evidence that this can be achieved in the improvement in mobility of the extrinsic carriers obtained for stretch oriented Durham polyacetylene over that for the unoriented films; along the direction of stretch this is as much as a factor of 1000 (Kahlert and Leising 1985, Townsend et al 1985). Secondly, towards the upper limit of charge concentration obtainable in the accumulation layer (up to 2×10^{13} charges/cm^2, see §2) it may be possible to move from the regime in which the isolated bipolaron hopping model we have used here is appropriate towards the 'metallic' regime in which screening by other charges present removes the barrier for transport. With chemical doping this occurs at between 0.1 and 1 molar % for polyacetylene . For the accumulation layer we are just on the lower side of this range. In order to see such a transition in mobility it will first be necessary to improve the structural order in the active layer of the polymer. As we discuss earlier in this paper, our spectroscopic measurements of the soliton excitations show clearly that for the case of devices which use silicon dioxide as the insulator layer we have a very high level of disorder.

6.2   Electro-optical properties

We have shown that the novel physics associated with conjugated polymers, that of formation of self-localised states which have states deep within the semiconductor gap, produces novel electro-optical properties. We are able to modulate optical coefficients deep within the semiconductor gap (at mid-gap for polyacetylene), in contrast to the situation for conventional semiconductors for which electro-optical modulation is usually close to the band edge where the absorption coefficient is high. In addition to the very convenient position in the spectrum for the optical modulation, conjugated polymers have the advantage of possessing very large electro-optical responses. The figure of merit in the present context is the optical cross-section, and for the mid-gap transition in polyacetylene this is of order 10^{-15} cm^2. This is a very large figure, and arises because the dipole matrix element for band to gap state transitions is enhanced over interband transitions by extent of delocalisation of the excitation on the chain, see for example Fesser et al (1983).

There is scope for a range of electro-optical devices based on polymers. One scheme to improve the coupling of the optical signal to the charge layer in the device is to couple the light within the device so that it runs parallel to the charge layer. This might be achieved by using either the insulator layer of the conjugated polymer layer (or both) as a waveguide for the optical signal.

REFERENCES

Assadi A Svensson C Willander M and Inganäs O 1988 *Appl. Phys. Lett.* **53** 195

Brown C S Vickers M E Foot P J S Billingham N C and Calvery P D 1986 *Polymer* **27** 1719

Burroughes J H Jones C A and Friend R H 1988 *Nature* **335** 137

Burroughes J H Jones C A and Friend R H 1989a *Synthetic Metals* **28** 735

Burroughes J H Friend R H and Allen P C 1989b *J. Phys.* D**22** 956

Burroughes J H Jones C A Lawrence R A and Friend R H 1990 NATO ARW *Conjugated Polymeric Materials: Opportunities in Electronics, Optoelectronics and Molecular Electronics* Mons, Belgium, *NATO-ASI Series E: Applied Sciences* **182**, 221 (Kluwer, Dordrecht).

Burroughes J H and Friend R H 1990 *Materials Research Society Fall Meeting, Boston, November 1989, MRS Symposium Proceedings* **173** 425

Colaneri N F Friend R H Schaffer H E and Heeger A J 1988 *Phys. Rev.* B**38** 3960

Chance R R Bredas J-L and Silbey R 1984 *Phys. Rev.* B**29** 4491

Ebisawa E Kurokawa T and Nara S 1983 *J. Appl. Phys.* **54** 3255

Edwards J H and Feast W J 1980 *Polymer Commun.* **21** 595

Edwards J H Feast W J and Bott D C 1984 *Polymer* **25** 395

Feast W J and Winter J N 1985 *J. Chem. Soc. Chem. Comm.* 202

Fesser K Bishop A R and Campbell D K 1983 *Phys. Rev.* B**27** 4804

Garnier F and Horowitz G 1987 *Synthetic Metals* **18** 693

Garnier F Horovitz G and Fichou D 1989 *Synthetic Metals* **28** 705

Grant P M Tani T Gill W D Krounbi M and Clarke T C 1980 *J. Appl. Phys.* **52** 869

Heeger A J 1986 in *Handbook on Conducting Polymers* ed. T. J. Skotheim, (Marcel Dekker, New York)

Inganäs O 1990 *STU Meeting on Conducting Polymers*, Gothenburg, Sweden.

Kahlert H and Leising G 1985 *Mol. Cryst. Liq. Cryst.* **117** 1

Kanicki J 1986 in *Handbook on Conducting Polymers* ed. T. J. Skotheim, (Marcel Dekker, New York), pp 544-660.

Kivelson S 1982 *Phys. Rev* B**25** 3798

Koezuka H Tsumura A and Ando T 1987 *Synthetic Metals* **18** 699

Koezuka H and Tsumara A 1989 *Synthetic Metals* **28** 753

Lawrence R A Burroughes J H and Friend R H 1989 *Springer Series on Solid State Sciences* **91** 127

Mott N F 1987 *Conduction in Non-Crystalline Materials* (Oxford University Press, Oxford)

Orenstein J 1986 in *Handbook on Conducting Polymers* T. J. Skotheim, ed. (Marcel Dekker, New York), pp 1297-1396

Rühe J Colaneri N F Bradley D D C Friend R H and Wegner G 1990 *J. Phys. CM* 2 5465

Su W-P Schrieffer J R and Heeger A J 1979 *Phys. Rev. Lett.* **42** 1698

Su W-P Schrieffer J R and Heeger A J 1980 *Phys. Rev.* B22 2099; Err. 1983 **B28** 1138

Sze S M *Physics of Semiconductor Devices* 1981 2nd Edition (Wiley-Interscience, New York).

Tomozawa H Braun D Phillips S Heeger A J and Kroemer H 1987 *Synthetic Metals* **22**, 63; *ibid.* 1989 **28** 687.

Townsend P D Bradley D D C Horton M E Pereira C M Friend R H Billingham N C Calvert P C Foot P J S Bott D C Chai C K Walker N S and Williams K P J 1985 *Springer Series in Solid State Sciences* **63** 50

Townsend P D and Friend R H 1989 *Phys. Rev.* B**40** 3112

Townsend P D and Friend R H 1987 *J. Phys. C* **20** 4221

Application of dopant-induced structure-property changes of conducting polymers

R.H. Baughman, L.W. Shacklette

Research and Technology, Allied-Signal Inc.
Morristown, New Jersey, USA

1. INTRODUCTION

The applications possibilities for change-transfer conducting polymers can be divided into two broad categories. The first category utilizes conducting polymers and environmental conditions in which the polymer redox state is effectively invariant. Included in this category are important applications in which conducting polymers replace either conventional conductors (metals, metal oxides, and carbons) or conventional insulating plastics - such as for conducting coatings and conducting composites applications. Recent advances in improving the environmental stability, processibility, and properties profiles of conducting polymers has resulted in major opportunities in this category - ranging from electromagnetic shielding to antistat applications.

The second category of applications is the focus of the present work. This category utilizes chemical or electrochemical redox for the operation of devices. Included in this category are: (1) high energy density batteries and redox capacitors, (2) electromechanical actuators, (3) electrochromic windows and displays, (4) electronically controlled membranes, and (5) sensor devices.

At least one doping, dedoping, or dopant compensation process is utilized in device operation for the second category of applications. The nature of these redox processes are described in Section 2, the associated property changes are examined in Section 3, and devices which utilize these property changes are discussed in Section 4.

2. OXIDATION AND REDUCTION PROCESSES USED IN DEVICE OPERATION

The nature of oxidation and reduction processes for conducting polymers is key for device applications. Specifically, the magnitude, rate, and reversibility of property changes are generally dependent upon the compositional changes during ion transfer processes and the associated structural changes. Inversely, both the chemical and structural nature of changes upon oxidation and reduction can strongly depend upon the oxidation and reduction rates. For example, depending upon the dopant concentration range, equilibrium structures result from either the incorporation of more dopant in pre-existing dopant-containing columns or planes or the creation of new dopant-containing columns or planes. However, since the latter processes are generally kinetically hindered by the required occurrence of nucleation events, high oxidation or reduction rates can result in nonequilibrium structures in which dopant concentration changes are preferentially accommodated by changes in the concentration of dopant in pre-exisiting dopant-containing planes or columns.

Furthermore, the chemical nature of oxidation or reduction processes can strongly depend upon the rates for these processes. For example, the rapid oxidation or reduction of the neutral polymer will tend to involve the counterion in an electrolyte bath which provides the most favorable kinetics for incorporation - rather than necessarily the most favorable thermodynamics. Specifically, oxidation of donor-doped polymer or the reduction of acceptor-doped polymer can be accompanied during high rate oxidation or reduction by the incorporation of acceptor or donor, respectively, rather than by the dedoping of donor or acceptor, respectively. Relevant to this point, Shimidzu et al. (1988) have shown that polymeric counter anions can be effectively immobile in conducting polymers, so that redox processes result from cation insertion-deinsertion. Again dependent upon oxidation and reduction rates and the solvents or electrolytes used for doping processes, solvent incorporation with dopant in the polymer can either occur or be excluded. Another complication is suggested by the work of Shinohara et al. (1986), who report that the size of the counterion used for electrochemical polymerization of a conducting polymer (polypyrrole) determines the kinetics of latter electrochemical redox with other ions. The claim is that an ion having a substantially larger effective radius will have a significantly decreased redox kinetics compared with that for an ion having a comparable or smaller effective radius than the initial counterion.

The kinetics of oxidation and reduction processes is important for many proposed applications of conducting polymers - such as in batteries, electrochromic displays, sensors, electromechanical actuators, and intelligent materials systems. Both device response rates and device cycle life are generally enhanced by minimizing dopant diffusion distances - such as by employing thin films or fibers. Similarly, the use of aqueous electrolytes is generally preferred over organic electrolytes for high rate devices because of the much higher obtainable conductivities for the former electrolytes. Since solid-state ion diffusion distances, the natures of diffusing ions and the conducting polymer, and electrolyte conductivities can profoundly effect redox rates, observed device response rates vary over an enormous range. For example, the times required for the full charge and discharge of conducting polymer batteries are typically of order an hour, both because of the usual use of thick electrodes (to increase energy density) and the use of organic electrolytes. On the other extreme, Lacroix et al. (1989) have reported electrochromic switching times down to 100 μs and current densities of up to 100 A/cm^2 for 1200 Å thick polyaniline in 2M sulfuric acid.

3.   PROPERTY CHANGES DURING OXIDATION AND REDUCTION

The principal property changes of interest for device applications are changes in electronic properties (especially, optical absorption spectra and electrical conductivity), ion diffusion rates, surface energies, redox potentials, mechanical moduli and strength, and physical dimensions.

The most dramatic property change upon doping is obtained for electrical conductivity, where more than fifteen orders of magnitude change is readily obtainable in going from the insulating, undoped polymer to the polymer charge-transfer complex. Changes in optical properties upon doping are similarly dramatic; typically a π-π * transition in the visible for the undoped polymer disappears and is replaced by an absorption peak or peaks at much lower energies. In some cases, the doped polymer has little absorption in the visible region, so that a transparent conductor results from doping. Except for devices whose operation depends upon complete switching between metal and insulator, cycling over the entire dopant concentration range is undesireable. The reason is several fold, the retention of appreciable electronic conductivity and the utilization of small changes in dopant concentration can provide optimized device response rates and high cycle lifetimes.

The dimensional changes upon doping and the effect of doping on mechanical properties is key for utilization of conducting polymers in electromechanical actuators and as property-changing materials for "intelligent systems". Additionally, these dimensional and mechanical property changes are important for many applications which require high cycle life, such as rechargeable batteries and electrochromic displays, since mechanical stresses due to inhomogeneous doping can result in materials failure.

Both changes in linear dimension and overall volume change on dopant insertion in a conducting polymer can strongly depend upon the dopant concentration range. This is a consequence of the typically observed packing arrangements in conducting polymer complexes. Dopant ions often form columns or planar assemblies which are inserted between the polymer chains. Depending upon the dopant concentration range, dopant can be accommodated by increased dopant density in preexisting dopant arrays (columns or sheets) or by the displacement of polymer chains during formation of an increased number of dopant arrays. The large dimensional changes corresponding to the compositional range of the latter process can be used in extensional electromechanical actuators. The small volume changes of the conducting polymer for the former process, coupled with large associated volume changes of counter electrode or electrolyte during redox, can be used in hydraulic electromechanical actuators. Polyacetylene doped with alkali metals provides important examples of both phase regimes (Baughman et al. 1985, Murthy et al. 1990, and Djurado et al. 1990). For example, conversion of a structure with four polymer chains per alkali metal column (y=0.0625 in CHK_y) to one with two polymer chains per alkali metal column (y=0.125) without change in intracolumn, interion separations results in a large volume expansion (12.5 cm^3/Faraday, which is about 27% of the molar volume of potassium). The corresponding percentage volume change for polyacetylene is 6.6%, or a 1.06% change in volume for a percent change in dopant concentration. This percentage volume change per percent change in dopant concentration (deduced from x-ray diffraction results) is close to that measured by bulk dimensional changes for sodium (1.0% by Francois et al. 1981) or for potassium (1% by Plichta 1989). In contrast, further increases in dopant concentration (from y= 0.125 to 0.167) results from a decreased interion separation in alkali metal columns, from 4.9 Å to 3.7 Å (Baughman et al. 1985 and Murthy et al. 1990). Only a small volume change is associated with this increased ion density in the alkali metal ion columns. Hence, the net volume change in this composition range on reduction of polyacetylene during oxidation of a potassium anode is close to the molar volume of potassium (45.9 cm^3/mole). For comparison with the above, the doping of polyacetylene with unsolvated lithium (up to y=0.11) produces a volume change (decrease) of the polyacetylene of only a few percent or less (Murthy et al. 1989). Consequently, the net volume change of a lithium anode and polyacetylene cathode during lithium doping is about -12 cm^3/Faraday (which is close to the molar volume of lithium).

Enormous dimensional changes can result from the doping of conducting polymers with larger dopants or with dopants which are solvated. For example, in-situ measurements of electrode buoyancy changes by Okabayashi et al. (1987) indicate that about 3 propylene carbonate molecules are reversibly inserted with each perchlorate ion during the oxidation of polyaniline in $LiClO_4$/propylene carbonate. The associated volume change of the polyaniline is 297 cm^3/Faraday - corresponding to a volume increase of the polyaniline by a factor of 2.2 over the total observed doping range. Despite the large volume increase of the polyaniline (compared with the initial volume of this polymer), the volume decrease of the electrodes and electrolyte is small (about -3 cm^3/Faraday). Using the results of in-situ bulk measurements of Slama and Tanguy (1989), the reversible volume change on oxidation of polypyrrole in propylene carbonate/$LiClO_4$ electrolyte has a similar value (272 cm^3/Faraday) as above discussed for polyaniline. This large value again results from solvent cointercalation with the anion.

For polymer backbones which are planar both before and after doping, the dopant-induced dimensional changes for this backbone are small. However, because of the high strength and modulus for the chain direction of highly oriented polymers, such dimensional changes can find application in electrochemical actuators. For example, the doping of polyacetylene with an electron acceptor causes chain-length expansion. For lithium, sodium, and potassium the maximum change (expansion) is from 1.0 to 1.6% and for iodine the maximum change (contraction) is about -0.4% (Murthy et al. 1987 and Winokur et al. 1988). While the percentage change in chain-axis length is a small fraction of the total percentage volume change for the larger dopants, this change provides a major contribution to the total volume change for polyacetylene doped with unsolvated lithium (Murthy et al. 1989). Also, because of a degree of chain misorientation and other disorder, the expansion in the orientation direction can exceed that deduced from x-ray diffraction measurements. The changes in chain-axis length of a conducting polymer can be quite large for polymers which change conformation as a consequence of doping, and can be comparable to the dimensional changes in orthogonal directions. However, while little change in per-chain modulus is expected for oriented polymers that do not change backbone conformation during doping, major decreases in this modulus can result if a transition occurs between planar and helical backbones.

Mechanical properties have not been optimized for many of the conducting polymers, and in notable cases these properties are limited by the low molecular weight of the polymer. Nevertheless, Akaji et al. (1989) reported extremely high modulus (100 GPa) and ultimate strength (900 MPa) for highly chain-oriented trans-polyacetylene obtained by a modification of the Naarman synthesis method. While it is easy to choose polymer/dopant combinations which result in poor mechanical properties for the doped polymer, examples are available which indicate that the mechanical properties of the doped polymer can be close to those of the undoped polymer, and in some cases exceed those of the undoped polymer. For example, Ito et al. (1988) found little change in Youngs modulus (2.6 GPa doped and 3.4 GPa undoped) or tensile strength (74 MPa doped and 81 MPa undoped) upon dedoping perchlorate-doped, unoriented polythiophene film. MacDiarmid et al. (1990) found that the Youngs modulus decreased from 8.6 GPa to 5.0 GPa and the ultimate tensile strength decreased from 366 MPa to 176 MPa upon doping drawn fibers of polyaniline (polyemeraldine base) with HCl. For comparison with these results, tensile strengths of 50-90 MPa were reported by Abe et al. (1989) for undrawn films of both undoped polyaniline and polyaniline doped with various protonic acids ($HClO_4$, HCl, H_2SO_4, and p-toluene sulfonic acid). Tokito et al. (1990) obtained extremely high mechanical properties both before and after iodine doping of oriented poly(2,5-dimethoxy-p-phenylene vinylene) fibers, combined with high conductivity for the doped state (1200 S/cm). The initial modulus of 35 GPa decreased to 25 GPa upon iodine doping, but the tensile strength (700 MPa) was essentially uneffected by doping. In contrast, Machado et al.(1989) obtained similarly high mechanical properties for undoped, oriented films of poly (p-phenylene vinylene), 37 GPa modulus and 500 MPa ultimate tensile strength, but these mechanical properties substantially degraded upon either AsF_5 or SbF_5 doping.

4.   APPLICATION OF PROPERTY CHANGES

4.1   Rechargeable Batteries and Redox Capacitors

The conducting polymer battery is among the many areas pioneered by MacDiarmid, Heeger, and coworkers (Nigrey et al. 1979). The first major commercial application of conducting polymers has been in button cell batteries of Bridgestone-Seiko (Nakajima and Kawagoe 1989). These rechargeable batteries utilize polyaniline as a positive electrode (cathode), lithium-aluminium alloy as the negative electrode (anode), and $LiBF_4$ in a mixture of propylene carbonate and 1,2-dimethoxyethane as electrolyte. During battery discharge, electrons move from the lithium alloy anode to the polyaniline cathode as Li^+

from the anode and BF_4^- from the cathode enter the electrolyte. As discussed in Section 3, the polyaniline undergoes a substantial volume change as the BF_4^- and coinserted solvent deinserts from the polyaniline. Since irreversible mechanical damage can result from resulting inhomogeneous electrode dimensional changes, battery cyclability substantially decreases with increasing charge/discharge rates and increasing depth of discharge per cycle . However, under relatively slow charge/discharge conditions and a 30% depth of discharge, a cycle life of 1000 cycles is reported. While this polymer battery is attractive for applications which require long cycle life, long float life, and low self-discharge rate, energy storage capacity is low, both for fundamental reasons and because of packaging in a button cell configuration with a large excess of the anode.

Any program to develop a high-storage-capacity, high-cycle-life polymer battery must deal with the following points. First, it is wasteful from a viewpoint of storage capacity to generate salts during battery discharge (as for the Bridgestone/Seiko battery) - since excess electrolyte volume for salt storage is required compared with the case when battery discharge results only in the movement of ions between electrodes. Second, inorganic cathode materials (such as Na_xCoO_2 and $Li_xVO_{2.17}$) are available which have high cycle life, high voltage (versus the alkali metal oxidation potential), and much higher volumetric storage capacity than is obtainable for cathodes consisting of acceptor-doped organic polymers. Third, alkali metals and alkali-metal alloys provide anodes having dramatically higher volumetric and gravimetric capacities than do alkali-metal-doped conducting polymers. However, the cyclability of alkali metals and alloys is poor and polymers (such as polyacetylene and poly(p-phenylene)) provide much higher cyclability.

The strategy which Allied-Signal utilized for the construction of high-energy-density, high-cycle-life batteries is to combine the differing strengths of the inorganic and conducting polymer electrode materials (Shacklette et al. 1987 and 1989). These batteries utilize inorganic ion inserting materials ($Li_xVO_{2.17}$ or Na_xCoO_2) for the cathode and an alkali metal alloy (Li_xPb or Na_xPb) with a conducting polymer (polyacetylene or poly(p-phenylene)) binder for the anode. The conducting polymer provides cyclability for the anode, rather than providing a significantly enhanced anode capacity. The conducting polymer binder for the alkali metal alloy forms a multiply connected, electronically and ionically conductive network within which the alloy particles are held. The fine fibrillar morphology of a conducting polymer holds the alloy particles as they become subdivided during cycling. The mixed ionic and electronic conductivity of the conducting polymer binder allows the alloy particles to continue the electronic and ionic processes associated with the charge and discharge of the cell, even if the alloy particles are completely surrounded by the conducting polymer. Additionally, the conducting polymer could serve to dissolve dendrites if they start to form. Consequently, high utilization is preserved for extended battery cycling.

Table I compares the gravimetric energy density of batteries which employ a conducting polymer, PPP, as a binder for an alkali metal alloy with the energy density of Ni/Cd batteries, Moli Energy lithium batteries, and polyaniline and polypyrrole batteries which utilize conducting polymers as the principle electroactive element in an electrode (Shacklette et al. 1989). Energy densities of 70 mWh/g have been achieved in packaged AF and AA size batteries, and energy densities of 100 mWh/gm are projected for the Na_xCoO_2 cathode battery because of the possibility of charging at higher potentials. The overall cycle lifetime is good and not limited in a balanced cell by the electrode containing the conducting polymer binder. Specifically, results obtained during exhaustive deep cycling of the Na-Pb/PPP binder anode provide an extrapolated capacity of 80% of the initial capacity after 1000 cycles.

TABLE 1   Comparison of Present and Future Performance of Packaged Cells[a]

Cell Couple	Average Potential	Demonstrated Energy Density (mWH/g)	Projected Energy Density (mWH/g)
Ni/Cd	1.2	39[b]	~45
Li_xMoS_2/Li	1.8	60-80[c]	80
$[(C_6H_4NH)(BF_4)_y]$/Li_xAl	2.5-3.0	4.4[d]	~40
$[(C_4H_3N)(BF_4)_y]$/Li	2.8	15[e]	~30
Na_xCoO_2/Na_xPb,PPP	2.5-2.8	65	70-100[f]
$Li_xVO_{2.17}$/Li_xPb,PPP	1.9	70	80

a	AF-or AA-size Metal can
b	SANYO High Energy "E-series" AF-size
c	Moli Energy, AA-size
d	Bridgestone/Seiko 2016 coin cell
e	AA-size BASF/VARTA
f	High value based on charge to 3.9V.

Conducting polymer redox capacitors, which function similarly to batteries, provide interesting applications opportunities. For example, Naegele (1989) has described a capacitor which utilizes polypyrrole as both anode and cathode. Shuttling of the chlorate ion between these two electrodes provides a capacitor-like dependence of voltage on charge. The interelectrode voltage is zero when both electrodes are charged to the same extent with the perchlorate ion. Deviation from this situation results in an interelectrode voltage which increases in proportion with interelectrode dopant transfer associated with electrochemical charging. Wrighton et al. (1985) have claimed conducting polymer capacitors of this type which have an energy storage density of 200-300 J/cm^3 (compared with about 1 J/cm^3 for conventional capacitors) and operating frequencies up to 100 Hz.

4.2   Electromechanical Actuators

Baughman et al. (1990) have recently proposed the application of conducting polymers for the direct conversion of electrical energy into mechanical energy. The concept is utilization of the large dimensional changes which occur upon either electrochemical donor or acceptor doping of polymers such as polyacetylene, poly(p-phenylene), polyaniline, polypyrrole, and polythiophene. As discussed in Section 3, these dimensional changes can provide a fractional length change ($\Delta L/L$) of over 10%, as compared with a maximum $\Delta L/L$ of about 0.1% at below depolarization voltages for the piezoelectric polymer poly(vinylidene fluoride). The much larger $\Delta L/L$ for conducting polymers, as compared with piezoelectric polymers, and the high mechanical strengths can provide a correspondingly increased work capacity per cycle. Such large dimensional changes of the conducting polymers are obtained using voltages which can be more than an order of magnitude lower than required for comparable electrostatic micromechanical actuators or for piezoelectric actuators.

Electrochemical mechanical actuators can generally function by using electrochemically induced (1) changes in a dimension of a conducting polymer, (2) changes in the relative dimensions of a conducting polymer and a counter electrode or, (3) changes in the total volume of a conducting polymer electrode, electrolyte, and counter electrode. In each case, a conducting polymer can serve as either only one electrode or as both electrodes.

Baughman et al. (1990) have previously described general features of microactuators which utilize either changes in linear dimension (extensional devices) or volume changes (hydraulic devices). Bimorph and unimorph actuators are special types of extensional actuators which are of particular interest, because such actuators provide high mechanical advantage and are most easily fabricated as microscopic devices.

A bimorph electromechanical cell can be designed analogously to well known bimorph structures for piezoelectric polymers. Unimorph and bimorph mechanical elements are herein defined according to the number of conducting polymer electrodes in the mechanical bender. A simple electrochemical bimorph cell consists of a polymer electrode strip and a polymer counter electrode strip cemented together by a polymeric electrolyte, which electronically separates these electrode elements. Alternately, the adhesive ion-conducting layer between electrodes can be a porous separator containing a liquid electrolyte. The major requirement for the operation of this bimorph cell is that the anode strip and cathode strip undergo differing changes in dimension upon passage of an electrochemical charge or discharge current. This is conveniently accomplished by using the same polymer as both anode and cathode strips and operating both polymer strips in the range of dopant concentrations that provides identical, but oppositely directed, transformations for anode and cathode strips during device operation. Optimal performance of the bimorph electromechanical cell will generally be obtained for cell designs in which the dopant shuttles between anode and cathode during operation, in contrast with designs where the dopant ions are stored in the solid-state electrolyte. Conducting polymer unimorph or bimorph actuators can be designed for applications on microcircuits. Possible applications include microtweezers, microvalves, micropositioners for microscopic optical elements, and actuators for micromechanical materials sorting (such as the sorting of biological cells). Strategies for fabricating conducting polymer micromechanical actuators can be based on techniques presently known for both the fabrication of micron dimensioned conducting polymer electronic devices and micromachined silicon mechanical devices.

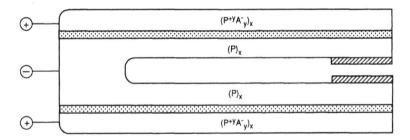

Figure 1.   Proposed use of paired bimorph actuators as microelectrochemical
               tweezers. Electrochemical transfer of dopant from the outer layer to
               the inner layer of each bimorph causes the opening of the tweezers.

Various microelectrochemical actuators, using either dimensional changes or volume changes, have been recently proposed by Baughman et al. (1990). One example is the use of conducting polymer bimorph flexors to form microtweezers, which is schematically illustrated in Figure 1. Each of the two bimorphs, which constitute the two arms of the microtweezers, can consist of layers of the same polymer which are separated electrically by an adhesive layer of solid-state electrolyte. The device operates by the electrochemical transfer of dopant between these conducting polymer layers. For example, reversing the electrode potentials shown in Figure 1 results in movement of the anion A^- from outer electrodes to inner electrode, so as to produce an opening of the tweezers. Chen et al. (1989) have fabricated electrostatic microtweezers which are 200 μm long and about 2.5 μm in the orthogonal dimensions. A major advantage of the proposed conducting polymer microtweezers is the low voltage required for operation, about 1 volt or less. For comparison, a voltage of over 100 volts was required for closure of the electrostatic microtweezers, which involved only about a degree change in the angle between the two arms of the tweezers.

The mechanical performance obtainable for conducting polymer electromechanical actuators can be calculated from the properties described in Section 3. For example, the maximum stress which can be developed in a reversible actuator by the electrochemical redox of a conducting polymer is the smaller of (1) the fractional change in dimension under zero load corresponding to a dopant concentration change Δy, $(\Delta L/L)_{0,\Delta y}$, multiplied by the Youngs modulus for the contracted state and (2) the limiting stress before mechanical failure, which is herein approximated as 50% of the ultimate tensile stress. Using the mechanical properties of undrawn polythiophene film, drawn polyaniline fibers, and drawn polyacetylene film, 50% of the ultimate tensile strength is reached under isometric conditions (fixed length conditions) for a $(\Delta L/L)_{0,\Delta y}$ of 1.1%, 2.1% and 0.45%, respectively. Consequently, only small changes in dopant concentrations are required under isometric conditions in order to develop high stresses (ca. 40, 180, and 450 MPa respectively). These stresses (corresponding to 380, 1900, and 4600 kgf/cm^2, respectively) are from one to two orders of magnitude higher than the tensile stress which can be developed by application of nondestructive voltages (ie, voltages which do not cause rapid depolarization) for the piezoelectric polymer poly(vinylidene fluoride). Specifically, using reported values (Lee and Marcus 1981) for the in-plane modulus (Y=3 GPa), the inverse piezoelectric constant ($d_{31} = 3 \times 10^{-11}$ m/V), and the maximum electric field which can be applied without rapid depolarization (ca. E= 3×10^7V/m for an alternating electric field), this stress for poly(vinylidene fluoride) is YEd_{31} or 2.7 MPa, compared with 40 to 450 MPa for the above mentioned conducting polymers. In addition to this major advantage of the conducting polymer actuator for high stress generation, the conducting polymer actuator has the significant advantage in requiring a much lower voltage for operation. Even for a film thickness as low as 1 μm, the above limiting field for the piezoelectric polymer corresponds to a voltage of 30 volts, while the conducting polymer electrochemical actuator would require a voltage of much less than a volt to generate the higher stresses. The above derived stress generation capabilities for conducting polymer electromechanical actuators are several orders of magnitude higher than observed (DeRossi et al. 1988) for isometric electromechanical contractions of salt-saturated polyacrylic acid/polyvinyl alcohol gels (ca. 3 kgf/cm^2 or 0.3 MPa), which are pH driven.

The mechanical work per polymer volume which can be accomplished in one electrochemical cycle provides another figure of merit which is impressive for properly designed conducting polymer electromechanical actuators. We consider here a tensile actuator operating under isotonic conditions (fixed mechanical load) and ignore changes in the elastic strain of the electromechanical polymer as a function of dopant level. The latter approximation will result in a serious overestimation of work density per cycle for a specified tensile load only when the product of polymer cross-sectional area and Youngs modulus is much lower for the contracted state than for the extended state, which is usually not the case. Using this approximation, the work density per cycle (involving a dopant concentration change Δy) is $\sigma(\Delta L/L)_{0,\Delta y}$, where σ is a stress below that required for irreversible deformation or fracture of the polymer. The value chosen for $(\Delta L/L)_{0,\Delta y}$ depends upon both the required cycle rate and cycle lifetime for the actuator, since both cycle rate and cycle lifetime generally decrease with increasing change in dopant concentration and increasing fractional dimension change. Consequently, we can conservatively assume a $(\Delta L/L)_{o,\Delta y}$ of at least a few percent for unoriented conducting polymers in high cycle life electromechanical actuators. Such a dimensional change is only about 10% or less of that available for complete doping of conducting polymers having large coulombic expansion coefficients. Hence, much higher work density per cycle could be achieved for actuators where high cycle life is not required. Since $\Delta L/L$ at below depolarization voltages for poly(vinylidene fluoride) is no larger than 0.1%, the conducting polymer electrochemical actuators designed for high cycle life could have more than an order of magnitude advantage compared with piezoelectric polymers in work density per cycle.

Among the advantages of the conducting polymers compared with piezoelectric polymers are more than order of magnitude increases for the achievable dimensional changes, the maximum electrically generated stress, and the maximum work density per cycle. Additionally, such performance can be achieved at voltages which can be about an order of magnitude lower than would be required for piezoelectric materials or, on the microscale, for electrostatic actuators. The major disadvantages of conducting polymers, compared with piezoelectric polymers, are provided by limitations on cycle life and cycle rate. Based on observed cycle lifetimes of conducting polymers in electrochemical optical displays, cycle lifetimes in excess of 10^6 cycles should be achievable in suitably designed actuators base on very thin films or fibers of conducting polymers. Additionally, by limiting the amount of charge transferred during the electrochemical cycle, such a cycle lifetime could perhaps be substantially exceeded. However, even under the best of circumstances, the cycle lifetime of the conducting polymer electrochemical actuator is much too low for use in motors which operate continuously at very high frequencies. Cycle times of about 100 ms should be feasible for conducting polymer microactuators, corresponding to the observed electrochemical switching times of thin conducting polymer films in electrochromic devices. Moreover, based upon the operation frequencies observed by Jones et al. (1987) for microelectrochemical transistors and Lacroix et al. (1989) for electrochromic displays, cycle times as short as 0.1 ms might be eventually achievable for very small microactuators.

Because cycle lifetime can be maximized and cycle time can be minimized by the use of very thin polymer films, conducting polymers will probably be of greatest interest for microactuators. Due to the likely prohibitive cost of using 10 micron or thinner films for larger actuators, and the absence of present technology for doing so, large scale actuators based on conducting polymers are likely to be usable only for applications which do not require either very high cycle life or very short cycle times. Examples of such applications are hydraulic or nonhydraulic actuators for window blinds or car door locks. Relevant for such applications, it is worthwhile noting that cycle lifetimes of about 10^3 cycles at 30% discharge and 10^4 cycles at a few percent discharge are claimed for nonaqueous electrolyte, polyaniline batteries manufactured by Bridgestone-Seiko (Nakajima et al. 1989). However, the use of a more highly conducting electrolyte and/or thinner electrodes would be required for the construction of an actuator with rate performance in the range of practical major interest.

4.3   Electrochromic Windows and Displays

Electrochromic displays is another interesting application area which utilizes the electrochemical doping and dedoping of conducting polymers. Depending upon the conducting polymer chosen, either the doped or undoped state can be either essentially colorless or intensely colored. However, the absorption of the doped state is dramatically red shifted from that of the undoped state. The color of this state can be modified using dopant ions which absorb in the visible (Skotheim et al. 1983). Because the conducting polymers typically have very high absorption coefficients in the visible (ca. 10^5 cm^{-1}) in at least the undoped or doped states, only very thin film coatings are required to provide display devices having high contrast and a very broad viewing angle. In contrast with the case for liquid crystal displays, the image formed by redox of a conducting polymer can have high stability even in the absence of an applied field.

Both rapid switching and reasonably high cycle lifetimes have been obtained for conducting polymer electrochromic devices. For example, using IR compensation to decrease the effect of electrolyte resistance, Lacroix et al. (1989) demonstrated electrochromic switching in less than 100 μs for 1200 Å thick polyaniline films in 2M sulfuric acid. Without using IR compensation, since this technique might not be practical for commercial display devices, Kobayachi et al. (1984) obtained high cycle life (about 10^6 cycles) and electrochromic switching between transparent yellow and green in less than 100 ms for 500 Å thick polyaniline films in 1 M HCl. Also, Foot and Simon (1989) obtained satisfactory switching after 10^6 cycles for both 2-ethoxy and 2-methoxy substituted polyaniline in 1 M HCl, as well as electrochromic switching times less than 2.5 ms for both oxidation and reduction.

Success in commercializing the electrochromic polymers for display applications would be enhanced by an increase in the cycle life to above 10^7 cycles. However, the present performance of these materials is quite attractive for use in electrochromic windows and mirrors for building and automotive applications. This is a promising applications area in which there is much industrial activity. For example, Allied-Signal researchers (Wolf et al. 1988) have described inventions in which conducting polymers are used as electrochromic materials for mirrors and windows - especially for the control of solar heating. Various useful embodiments are described, such as the use of anodes and cathodes which both provide color changes; the use of a chain-oriented, polarizing conducting polymer electrode to reduce glare; the use of two conducting polymer electrodes with crossed polarization directions to increase absorption efficiency; and the use of highly conducting polymeric electrolytes, such as a mixture of phosphoric acid and poly(vinyl alcohol). By proper choice of conducting polymer and polymer thickness, high transparency in the visible, high absorption in the solar infrared, and high transparency in the ambient temperature infrared is obtained. Hence, it is possible to optimally design windows to achieve desired aesthetic and solar energy management goals.

Also relevant for the window application, where rapid device response is not needed, Yoshida et al. (1988) of Toyota have demonstrated an electrochromic cell which provided about 10^6 cycles and switching times of about a second for an optical transmissivity change from 80% to 30%. Both electrodes in this organic liquid electrolyte cell are electrochromic: a WO$_3$ electrode which changes from colorless to blue upon reduction and a polyaniline electrode which changes from light yellow to blue upon oxidation. Consequently, the net color change is transparent light yellow to blue. Also, Akhtar et al. (1988) of Chronar Corporation have obtained response times shorter than a second, cycle lifetimes of many thousand cycles, and good color contrast for polyaniline electrochromic cells which use sulfuric acid/poly(ethyleneimine) as solid state electrolyte. An unusual method for electrode formation was used instead of the normally used electropolymerization route, since the polyaniline film was deposited by sublimation of chemically prepared polyaniline base or polyaniline hydrochloride.

Inganas and Lundstrum (1984) and Yoneyama et al. (1986) have used the electrochromism of conducting polymers to make photoelectrochromic memory and display devices. Instead of using transparent metallic electrodes, as in the above-described electrochromic devices, these photoelectrochromic devices utilize a conducting polymer (polypyrrole, poly(N-methylpyrrole, or polyaniline) deposited directly or indirectly on n-type silicon. These polymers change color at cathodic potentials and return to the original color at anodic potentials, but both of these write and erase steps do not occur in the absence of photogenerated carriers in the n-type silicon. Hence, convenient means are provided for the writing and erasure of information using a light beam. The time required for write and erase steps (100 ms for polyaniline) must be decreased and cycle life needs to be increased before this technology can be used for a high performance information storage system. Yoshino et al. (1985) had previously shown that conducting polymers like polythiophene can be used as optical memory elements. These authors utilized the photoinduced doping of polythiophene films containing diaryliodonium salts to provide nonelectrochemical memory elements which permits light-induced writing, but not light-induced erasure.

4.4 Electrochemically Controlled Chemical Separation and Delivery Systems

The redox dependent permeability and ion storage capability of conducting polymers provides a number of applications possibilities. One specific example is conducting polymer membranes having electrochemically controlled ion permeability, gas permeability, or molecular size selectivity. Early work by Burgmayer and Murray (1984) showed that electrochemical oxidation can be used to dynamically and reversibly change the anionic permeability of polypyrrole membranes by more than 1000 fold. Such permeability changes associated with a redox reaction might be used either in separation technologies or in the controlled storage and release of chemicals.

The ion storage capacity of the redox conducting polymers is also of interest for both chemical separation and chemical storage and release. A novel approach for using polypyrrole as both an anion-exchangeable and a cation-exchangeable electrode has been demonstrated by Shimidzu et al. (1988) for the deionization of water. The trick employed was to utilize two polypyrrole electrodes, the first electrochemically synthesized with an immobile polymeric anion (polyvinylsulfonate) and the second polymerized with a highly mobile anion such as Cl⁻, which was electrochemically deinserted during reduction. During the deionization process (1) electrochemical reduction of the electrode containing immobile polymeric anions occurs by cation insertion and (2) electrochemical oxidation of the counter polypyrrole electrode occurs by anion insertion. During the regeneration step, these processes are reversed electrochemically.

Numerous other possibilities exist for the use of conducting polymers in separation technologies. For example, the work of Garnier (1989) suggests that chiral conducting polymers can be used for the separation of chiral anionic species in racemic mixtures, since the relative handedness of chiral substituents on the conducting polymer and a chiral anion effects the rate of anion incorporation.

Relevant to the use of conducting polymers for chemical or drug release, Zinger and Miller (1984) have provided some interesting results. These authors demonstrated that a glutamate (a neurotransmitter) can be electrochemically "loaded" into polypyrrole films during the oxidation of polypyrrole in sodium glutamate solution. Glutamate release into an aqueous NaCl solution occurred during the reduction of the polypyrrole. Based on this work, one can imagine the application of conducting polymers as drug release materials (applied to the skin or implanted in the body) which would provide a drug at a release rate determined by feedback from sensors for the human body. Shinohara et al. (1985) have already demonstrated reversible electrochemical storage and release of glutamic acid from pyrrole microelectrodes similar in design to those which might eventually be used for release of this neurotransmitter in medical applications. Another novel application of conducting polymers as biochemical release materials is provided by Shimidzu (1987).This author described the use of polypyrrole on a polymeric support for the synthesis of either homo-oligoribonucleotides or sequence-defined oligoribonucleotides. The polypyrrole anchors the oligoribonucleotide until synthesis is completed, at which point, electrochemical reduction of the polypyrrole releases this material.

Kossmehl (1990) has demonstrated a quite different approach for the use of redox reaction of conducting polymers for the selective release of chemicals. Instead of storing chemicals as dopants in the conducting polymer and releasing them by electrochemical reduction of this doped polymer, Kossmehl (1990) used electrochemical redox of a conducting polymer for changing the wettability of the polymer surface. The specific application targeted is a printing press in which the text can be continuously changed. A matrix of electrodes is used to control the redox state of individual "dots" on a polythiophene-coated ink transfer cylinder. Depending upon whether the polythiophene dot is hydrophilic (i.e., doped) or hydrophobic (i.e., updoped), ink is either picked up by the dot or not. Transfer of this ink to paper provides the printed image, which can be continuously changed electrochemically, as needed, to provide new text.

4.5   Indicators and Sensors

Novel indicator devices can be constructed whose operation results from the dramatic changes in electrical conductivity of conducting polymers upon redox reaction. Baughman et al. (1987) have described a generally applicable technology which utilizes commercially-available radio frequency anti-thief targets, costing about $0.03 each, for the construction of inexpensive, remotely readable, in-box sensors for time-temperature history, temperature, temperature limits (freeze and defrost indicators), irradiation dosage, and mechanical shock. The anti-theft target is a centimeter-dimensioned radio frequency antenna. In a typical device design, a conducting polymer layer is combined with layers containing either doping agents or dopant compensating agents. The indicator device response results from the release of doping agent or dopant compensating agent in response to a desired ambient influence, so as to provide a conductivity change of the conducting polymer. Depending upon the conductivity of the conducting polymer, the antennae response characteristics change (via direct shorting of the antenna, capacitance coupling, or radio frequency shielding). This charged antennae response, as consequence of a specified ambient variable being monitored, is measured remotely using a hand-held radio frequency transmitter/receiver. Consequently, without opening a shipping container which contains both a perishable product and the indicator device, various ambient influences which effect product quality can be conveniently and inexpensively monitored.

Among the various chemical sensors proposed using conducting polymers, biological sensors are especially interesting. For example, Shimidzu (1987) and coworkers showed that adenosine triphosphate, poly(adenylic acid), uridine triphosphate, poly(uridylic acid), or a sequence-defined oligonucleotid could be incorporated into polypyrrole during electrochemical polymerization. Sensor function was demonstrated by measurement of changes in the surface potential of the nucleotide-doped polypyrrole dependent upon the nucleotide contained in test solutions.

Iwakurra et al. (1988) demonstrated the operation of a simple biosensor for glucose concentration. This sensor is based upon polypyrrole electrochemically polymerized on an electrode from a solution which includes both glucose oxidase and ferrocenecarboxylic acid, which is a mediator for electron transfer. Glucose concentration was determined amperometrically by measurement of steady state current when a fixed potential was applied to the electrodes. This steady state current, which was an monotonically increasing function of glucose concentration, was achieved within a few minutes because of the presence of the entrapped mediator in the polypyrrole.

Conducting polymer electrode sensors have been miniaturized so as to provide micron spatial resolution in concentration measurements and microsecond scale time response. General discussion of such activities are available from Wightman (1988), Chidsey and Murray (1986), and Thackeray et al. (1985). Measurement of redox currents due to electrode or electrolyte reactions or changes in electrode potential are commonly used to provide sensor response. However, changes in the resistivity of a conducting polymer (as a consequence of redox reaction with the chemical being analyzed) provides a particularly sensitive method of analysis. This sensitivity results from the many orders of magnitude change in conductivity which results during conducting polymer redox.

Using a conductivity change of poly(3-methylthiophene) in the electrochemical analogue of a transistor, Thackeray et al. (1985) demonstrated sensor response to $< 10^{-15}$ mole of an oxidant. A schematic representation of a electrochemical sensor analogous to a transistor is shown in Figure 2. Depending upon the concentration of dopant ion present in the electrolyte, the gate potential V_g provides a "gate current" I_g between the conducting polymer and the counter electrode (gate electrode). The corresponding redox reaction of the conducting polymer dramatically changes the conductivity of this polymer, which is measured by the drain current (I_d) resulting from a given drain voltage (V_d). In the form of microelectrochemical transistors, device operation at switching frequencies exceeding 10 kHz (Jones et al. 1987) and resistivity changes per switching cycle above a factor of 10^5 (Thackeray et al. 1985) has been demonstrated. Using photolithography and shadow deposition methods, Jones et al. (1987) have reduced the drain-source separation to 500Å and obtained a total width of 8μ for the drain-source array (which is 50 μ long). As a consequence of such short drain-source separation and the correspondingly small amount of conducting polymer (10^{-14} mole of electrochemically deposited polyaniline) required to bridge the drain-source gap, device response times of less than 100μs were obtained, along with device response to less than 10^{-10} coulombs of charge.

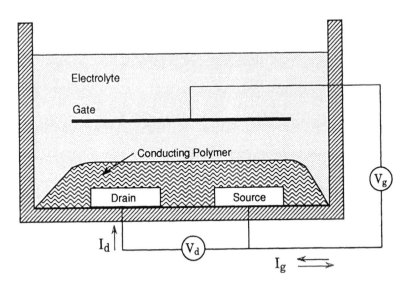

Figure 2.   Schematic representation of a electrochemical sensor analogous to a transistor

5.   CONCLUSION

The enormous applications possibilities for conducting polymers are only now beginning to be commercially realized. High volume manufacture will eventually be required for applications in which conducting polymers replace conventional metals, plastics, and composites. Only small amounts of polymer are needed for the presently described redox device applications - but conducting polymers provide the key element which provides function for high value devices. As such, this category of applications is well suited for early commercialization, as already demonstrated by the commercialization of polymer batteries by Bridgestone-Seiko. The performance of prior art materials having the benefit of long R & D efforts provides challenging benchmarks for conducting polymer devices. However, no alternate materials provide the novel property combinations of conducting polymers or the same ability to dynamically change properties on command.

REFERENCES

Abe M, Ohtani A, Umemoto Y, Akizuku S, Ezoe M, Higuchi H, Nakamoto K, Okuno A and Noda Y 1989 *J. Chem. Soc. Chem Commun* 1736
Akaji K, Soezaki M, Shirakawa H, Kyotani H, Shimomura M and Tanabe Y 1989 *Synthetic Metals* **28** D1
Akhtar M, Weakliem H A, Paiste R M and Gaughan K 1988 *Synthetic Metals* **26** 203
Baughman R H, Shacklette L W, Murthy N S, Miller G G and Elsenbaumer RL 1985 *Mol. Cryst. Liq. Cryst.* **118** 253
Baughman R H, Elsenbaumer R L, Iqbal Z, Miller G G and Eckhardt H 1987 *Electronic Properties of Conjugated Polymers* eds H Kuzmany, M Mehring and S Roth (Berlin: Springer-Verlag) pp 432-39
Baughman R H, Shacklette L W, Elsenbaumer R L, Plichta E and Becht C 1990 *Conjugated Polymeric Materials: Opportunities in Electronics, Optoelectronics, and Molecular Electronics* eds J L Bredas and R R Chance (Netherlands: Kluwer) pp 559-82
Burgmayer P and Murray R W 1984 *J. Phys. Chem.* **88** 2515
Chen L Y, Zhang Z L, Yao J J, Thomas D C and MacDonald N C (1989) *Micro Electro Mechanical Systems, Proceedings of IEEE* eds S C Jacobsen and K E Petersen pp 82-7
Chidsey E D and Murray R W 1986 *Science* **231** 25
DeRossi D, Domenici C and Chiarelli P 1988 *Sensors and Sensory Systems for Advanced Robotics* ed P Dario (Berlin: Springer-Verlag) pp 201-18
Djurado D, Fischer J E, Heiney P A, Ma J, Coustel N and Bernier P 1990 *Synthetic Metals* **34** 683
Foot P J S and Simon R 1989 *J. Phys. D: Appl. Phys.* **22** 1598
Francois B, Mermilliod N and Zuppiroli L 1981 *Synthetic Metals* **4** 131
Garnier F 1989 *Angew. Chem. Int. Ed. Engl* **28** 513
Ingonas O and Lundstrom I 1984 *J Electrochem. Soc: Solid-State Science and Technology* **131** 1129
Ito M, Tsurono A, Osawa S and Tanaka K 1988 *Polymer* **29** 1161
Iwakura C, Kajiya Y and Yoneyama H 1988 *J. Chem. Soc. Chem Commun.* 1019
Jones E T T, Chyan O M and Wrighton M S 1987 *J. Am. Chem. Soc.* **109** 5526
Kobayashi T, Yonegama H and Tamura H 1984 *J. Electroanal. Chem.* **161** 419
Kossmehl G 1990 Unpublished
LaCroix J C, Kanazawa K K and Diaz A F 1989 *J. Electrochem. Soc.* **136** 1308
Lee, J K and Marcus M A 1981 *Ferroelectrics* **32** 93
MacDiarmid A G 1990 Private communication
Machado J M, Masse M A and Karasz F E 1989 *Polymer* **30** 1992
Murthy N S, Shacklette L W and Baughman R H 1987 *J. Chem Phys.* **87** 2346

Murthy N S, Shacklette L W and Baughman R H 1989 *Phys. Rev.* B **40** 12550

Murthy N S, Shacklette, L W and Baughman R H 1990 *Phys. Rev.* B **41** 3708

Naegele D *1989 Electronic Properties of Conjugated Polymers III* eds H Kuzmany, H Mehring and S Roth (Berlin: Springer-Verlag) pp 428-431

Nakajima T and Kawagoe T 1989 *Synthetic Metals* **28** C629

Nigrey P J, MacDiarmid A G and Heeger A J 1979 *J. Chem. Soc., Chem Comm.* 594

Okabayashi K, Goto F, Abe K and Yoshida T 1987 *Synthetic Metals* **18** 365

Plichta E 1989 Masters Thesis, Rutgers Univ., New Brunswick, New Jersey

Shacklette L W, Maxfield M, Gould S, Wolf J F, Jow T R and Baughman R H 1987 *Synthetic Metals* **18** 611

Shacklette L W, Jow T R, Maxfield M and Hatami R 1989 *Synthetic Metals* **28** C655

Shimidzu T 1987 *Reactive Polymers* **6** 221

Shimidzu T, Ohtari A and Honda K 1988 *J. Electroanal. Chem.* **251** 323

Shinohara H, Aizawa M and Shirakawa H 1985 *Chemistry Letters* 179

Shinohara H, Aizawa M and Shirakawa H 1986 *J. Chem. Soc., Chem.Commun.* 87

Skotheim T A, O'Grady W E and Linkous C A 1983 (patent filing),US Patent 4,571,029

Slama M and Tanguy J 1989 *Synthetic Metals* **28** C171

Thackeray J W, White H S and Wrighton M S 1985 *J. Phys. Chem* **89** 5133

Tokito S, Smith P and Heeger A J 1990 *Polymer*, in press

Wightman R M 1988 *Science* **240** 415

Winokur M J, Moon Y B, Heeger A J, Barker J and Bott D C *Solid State Commun.* **68** 1055

Wolf J F, Miller G G, Shacklette L W, Elsenbaumer R L and Baughman R H 1988 (patent filing), US Patent 4,893,908

Wrighton M S, White H S and Thackeray J W 1985 (patent filing), US Patent 4,417,673

Yoneyama H, Wakamoto K and Tamura H 1986 *Materials Chemistry and Physics* **15** 517

Yoshida T, Okabayaski K, Asaoka T and Katsushi A 1988 *Electrochemical Society Extended Abstracts* **88-2** 552

Yoshino K, Sugimoto R, Rabe J G and Schmidt W F 1985 *Japanese Journal of Applied Physics* **24** L33

Zinger B and Miller L L 1984 *J. Am. Chem. Soc.* **100** 6861

Applications of conducting polymers in computer manufacturing

Marie Angelopoulos, Jane M. Shaw and John J. Ritsko

IBM T.J. Watson Research Center, Yorktown Heights, NY 10598

U.S.A.

ABSTRACT

By the addition of onium salts, the emeraldine base polyaniline has been made radiation sensitive and shown to exhibit the qualities of a high resolution single layer resist for integrated circuit lithography. Thin conducting films of polyaniline are also shown to be effective electron beam discharge layers when incorporated into a multilayer electron beam resist system.

1. INTRODUCTION

While it is unlikely that conducting polymers will be used as active devices or even as metallic conductors in integrated circuits or their interconnection packages, there is a good chance that they will fill important roles in the manufacturing processes for these components. As the active devices on computer chips become smaller with more and more circuits on a chip, the materials and processes used to make these chips continue to evolve toward more and more sophisticated structures.   The intricate patterns of the variously doped regions of silicon on a chip and their metallic interconnections are formed by the methods of lithography (Greek: writing on stone). In the lithographic process a radiation sensitive material, called a resist, which is usually a polymer approximately 1 micron thick, is coated on the silicon wafer and exposed to radiation in a particular pattern (Shaw 1988). This image is then developed leaving some areas of silicon covered by the polymer and others which are now open and which can be further processed. The applications of conducting polymers described in this paper have to do with improved resist processes. The polymer is polyaniline, and the features of the material which make it desirable are the difference in solubility between doped and undoped states and its relatively high conductivity in very thin layers. Polyaniline can be made to be photosensitive and act as a resist.

2. FET FABRICATION

As an example of resist processes, a simplified method for making a field effect transistor (FET) is described in crossection in figure 1. A thin oxide (a few hundred angstroms thick) is grown by thermal processes on a silicon wafer. Polycrystalline silicon is sputter deposited on this wafer and a thin layer of photosensitive or electron beam sensitive polymer is blanket coated on the top. A thin stripe of resist for each FET is made by exposure to the radiation. The radiation induces a difference in solubility between exposed and unexposed regions, development in the solvent leaves the desired pattern on the wafer. This resist stripe will define the most critical feature of the FET namely the length of the channel between the source and drain. Once the resist is patterned, the wafer, which contains on the order of a hundred million of these FETs is subjected to a plasma which etches away the polycrystalline silicon and (with different gases in the plasma) the silicon oxide. The plasma does not etch the resist. Then the source and drain are made by heavily doping with blanket ion implantation. The ions do not affect the resist. Finally the resist is stripped from the gate and high conductivity metallic contacts are formed over the source, gate and drain by reaction with organometallic gases to form metal silicides.

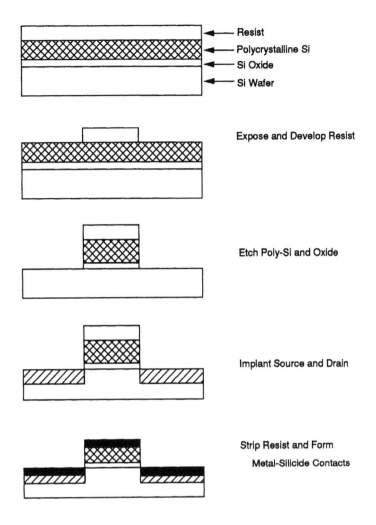

Figure 1. Field Effect Transistor (FET) Fabrication

In the commercially available one megabit and four megabit memory chips the active devices are FETs with channel lengths of 1 micron and 0.7 microns respectively. IBM has recently announced that it has fabricated on a manufacturing line 16 megabit memory chips with 0.5 micron channel length. Each chip has an area of roughly 1 cm² and a single 200 mm wafer contains several hundred chips. Nevertheless, neither the limits of

integration nor the smallness of devices have yet been reached
in commercial applications.   Rishton has recently shown that
advanced electron-beam lithographic techniques can be used to
make functioning FETs with 700 A channel lengths (Rishton 1988,
Kern 1988)

3.   SINGLE LAYER RESISTS

A single layer resist is a polymer which is coated directly on
the material to be patterned and one in which irradiation creates a solubility difference between exposed and unexposed regions.   Polyaniline (shown in figure 2a) in the emeraldine base
oxidation state (y= 0.5) is non-conducting and soluble in N-
methylpyrrolidinone (NMP).   But, once doped into the conducting
state polyaniline (emeraldine salt) is not soluble in NMP.   Thin
films of the polymer in the emeraldine oxidation state can be
converted to several conducting salts including emeraldine
hydrochloride ($\sim$ 1 ohm^{-1}cm^{-1}), emeraldine base hydrosulfonate
($\sim$ 1 ohm^{-1}cm^{-1}), and emeraldine hydroacetate ($\sim$ 0.1 ohm^{-1}cm^{-1}),
by soaking for 6 hours in the appropriate dilute aqueous acid
solution, 1N HCl, 1N H_2SO_4, and 80% acetic acid respectively
(Angelopoulos 1989, 1990).   In order to achieve a radiation

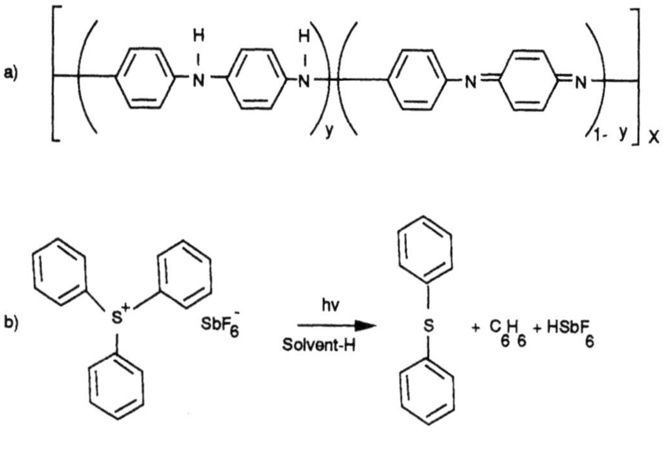

Figure 2. a) Polyaniline, b) Onium Salt Photolysis

induced insolubility, the emeraldine base must be doped with materials which produce protonic acids when irradiated. One such class of materials are onium salts (Crivello 1977, 1979). To make a radiation sensitive polymer triphenylsulfonium hexafluoroantimonate (figure 2b) was dissolved in NMP and mixed with the emeraldine base (also in an NMP solution). The resulting two component solution was used to spin coat thin films on quartz and silicon wafers. The samples were then exposed to ultraviolet radiation (240nm) or to an electron beam system. When exposed to radiation (100-300 mJ/cm^2 at 240 nm) the original blue film turns green characteristic of the conducting state (Angelopoulos 1990). Conductivities as high as 0.1 ohm^{-1}cm^{-1} have been measured. When the unexposed material is washed away in NMP, conducting lines as small as 0.5μm could be generated. While further development will be required to obtain even higher resolution, polyaniline appears to be the most promising conducting polymer resist at this time.

4. MULTILAYER RESISTS

In some applications single layer resists are inadequate. The topography of thick wiring layers on chips would require a near UV optical exposure system to have an impossibly long depth of focus to properly expose a single layer resist which covered the topography conformally. If the resist did not conformally cover the topography, its thickness would vary over the substrate, and features defined in the resist would have varying widths after development. In electron-beam systems the incident electrons penetrate a thick single layer resist and create secondary electrons in the wafer which are emitted in all directions. Some of these electrons are emitted into the resist and broaden the exposed regions reducing the resolution. The solution to these problems is to make a multilayer resist consisting of a thick radiation insensitive polymer which produces a smooth surface even over topography, with a thin radiation

sensitive resist on top. In electron-beam lithography the thick
resist absorbs the backscattered secondary electrons, thus improving the resolution. After the thin resist is exposed and
developed the pattern is transferred through the thicker lower
polymer by reactive ion etching.

While electron beam lithography is capable of extreme precision, the position of the beam is affected by charging of the
resist system during e-beam writing. This can cause pattern
displacement and misregistration. Thick multilayer resists
trap charge and are not feasible for dense pattern writing unless a conducting discharge layer is incorporated into the resist stack. An effective discharge layer is a thin film of
conducting polyaniline placed between the thick planarizing
polymer and the radiation sensitive resist (Angelopoulos 1989).

The multilayer resist used in these studies was made of a
2.8μm planarizing underlayer coated with 2000Å of non-
conducting emeraldine base by spin coating from NMP solution
followed by baking at 90°C for 15 minutes. The emeraldine base
was converted to several conducting salts by soaking in dilute
aqueous acid solutions as described previously. In one experiment the emeraldine base was treated with HCl of appropriate
pH to achieve different doping levels and different levels of
conductivity (Angelopoulos 1989). After doping and baking in
vacuum to remove water, a 1.2μm film of positive AZ type resist
was coated over the polyaniline.

The multilayer resist stacks were then subjected to an e-beam
charging test to measure the effectiveness of the discharge
layers. The samples were exposed at a dose of 15 μC/cm^2 at 25Kev
which is typical for many e-beam resists. Initially a 20 x 20
matrix of 2μm squares spaced 250μm apart was written on a 5mm
chip. Because the pattern is so sparce charging is negligible.

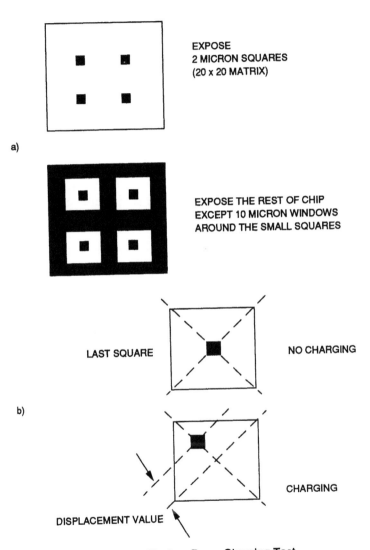

Figure 3. Electron Beam Charging Test

The chip is then completely overwritten except for $10\mu m$ by $10\mu m$ windows centered over each inner $2\mu m$ square as shown in figure 3a. Following exposure the patterns are developed and evaluated with a microscope to determine the degree of pattern displacement which is related to the amount of charging during exposure. When no charging occurs the $2\mu m$ squares will be located at the center of the $10\mu m$ windows as shown in figure 3b. Due to charging during the writing of the dense pattern the pattern will become offset as the charge builds up. The first window written will be least affected by charging because only a small area has been exposed. However, the last window (400th) is strongly influenced by charging because of the large surrounding area which has been exposed to the beam.

If no discharge layer or a non-conducting emeraldine base interlayer is used, the $2\mu m$ square can be offset by as much as $5\mu m$ from the center of the $10\mu m$ window for the last window written. However discharge layers of emeraldine hydrosulfonate and emeraldine hydrochloride (~ 1 ohm^{-1}cm^{-1}) as well as emeraldine hydroacetate (0.1 ohm^{-1}cm^{-1}) showed no measurable pattern displacement.

In order to establish the minimum conductivity required for zero pattern displacement on the e-beam discharge test, a series of interlayers with different doping levels were evaluated. Table I shows the pattern displacements which were observed for five partly protonated emeraldine base interlayers. From this table it can be seen that a conductivity of at least 10^{-4}ohm^{-1}cm^{-1} is required for the emeraldine salts in order to serve as successful discharge interlayers in the particular multilevel resist stacks described here. Below this value unacceptable pattern displacements occur during the e-beam writing process.

TABLE I. The dependence of the pattern displacement on the
          conductivity of the partly protonated emeraldine
          discharge interlayers.

pH of HCl Solution	Conductivity ($ohm^{-1}cm^{-1}$)	Pattern Displacement Observed (μm)
2	0.1	0
2.5	5.0×10^{-4}	0
2.6	3.0×10^{-4}	~<1
2.9	8.0×10^{-6}	~2
3.1	$<10^{-6}$	~>2

The photosensitive polyaniline described in the section on
single layer resists has a conductivity of ~0.1 $ohm^{-1}cm^{-1}$ upon
exposure to UV radiation. It too can be a good discharge layer
for multilayer resist systems. The photosensitive polyaniline
has the advantage that once cast as a thin film it can be very
quickly made conducting by a short blanket exposure to UV light
whereas the non-conducting emeraldine base films require several hours of soaking in acid solution to become conducting.

References:

Rishton S.A., Schmid H., Kern D.P., Luhn H.E.,Chang T.H.P., Sai-Halasz G.A., Wordeman M.R., Gamin E., and Polcari M., 1988 J. Vac. Sci. Tech. B6 140

Kern D.P., Rishton S.A., Chang T.H.P., Sai-Halasz G.A., Wordeman M.R., and Ganin E. 1988 J. Vac. Sci. Tech. B6 1836

Angelopoulos M., Shaw J.M., Kaplan R.D., and Perrault S. 1989 J. Vac Sci Tech B7 1519

Shaw J. M., 1989, Imaging Processes and Materials, Neblette's Eighth Ed. Van Nostrand Reinhold, New York, pp 567-586

Angelopoulos M., Shaw J.M., Huang W.S., Kaplan R.D. 1990, Molec. Cryst. Liq. Cryst, In Press

Crivello J.V. and Lam J.H.W. 1977, Macromolec.10, 1307

Crivello J.V. and Lam J.H.W. 1979, J. Poly. Sci. Poly. Chem. 17, 977

Conjugated oligomers as active materials for electronics

F. GARNIER, G. HOROWITZ, D. FICHOU and X. PENG
Laboratoire des Matériaux Moléculaires, CNRS, 2 rue Dunant,
94320 Thiais France

ABSTRACT: Metal-insulator-semiconductor field-effect transistors, MISFET, were fabricated from vacuum evaporated conjugated α-sexithienyl, α-6T, and characterized. As other organic based FET's, these devices operate through the injection of majority carriers in an accumulation channel. After a 3 hour heat treatment at 120°C, the obtained field effect mobility reaches the value of 10^{-3} $cm^2V^{-1}s^{-1}$, the highest reported one for an organic semiconductor. The characteristics of these devices can be further improved by reducing the ohmic current flowing in the bulk of the semiconductor, leading to thin film transistors.

1. INTRODUCTION

Conjugated oligomers have been the subject of a large number of studies, which have been mainly devoted to their conducting and electroactive properties. Much less work has been reported on their semiconducting properties, which allow them to be used as active components in electronic devices such as Schottky diodes[1-3] and metal-insulator-semiconductor field-effect transistors, MIFET's[4-7]. The characteristics of these devices are largely limited by the poor semiconducting properties of these polymers. Thus their carrier mobilty μ, limted by traps and defects to a value of about 10^{-4} $cm^2V^{-1}s^{-1}$, appears considerably lower by a power of 3 to 4 as compared to amorphous silicon. We have developed in our laboratory a new class of easily processable organic semiconductors, the conjugated oligomers, which, besides of representing interesting models for their corresponding polymers, show largely enhanced semiconducting properties as will be presented in this paper describing the characteristic of field-effect transistors fabricated from α-6T. We will also show that the use of a simple model of operation of MISFET's allows the improvement of their characteristics, through the realization of thin film devices.

2.   EXPERIMENTAL

The chemical synthesis of alpha-conjugated sexithienyl will be described in a forthcoming paper[8]. The schematic view of the device used in this study is given in Figure 1. A n-type silicon wafer (0.002 Ωcm^{-1}, <100> axis, 0.5 mm thick) was covered with a thermally grown oxide film (150 to 200 nm thick) and used as substrate on which α-6T was vacuum evaporated at a pressure of 5×10^{-3} Pa. Two gold contacts were then evaporated through a mask on top of the organic layer (70 nm thick, 310 µm width, separated by 90 µm), forming the source and drain contacts of the FET device. The back side of the Si substrate was cleaned from SiO$_2$ and covered with In-Ga alloy to form the gate contact.

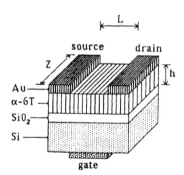

Fig. 1. Three-dimensional view of a field-effect transistor, W : channel width, L : channel length.

The thickness of the α-6T layer has been measured from the absorption spectrum of the same layer deposited on a glass slide under the same conditions. The electrical measurements were per-formed at room temperature, under ambient atmosphere, in an

electrically shielded box. Contacts were taken on the gold source
and drain contacts using tungsten microprobes (Micromanipulator).
A Hewlett Packard 4140 B picoammeter/d.C. voltage source, monitored with a HP 310 desk computer, was used for measuring the
FET characteristics.

3. RESULTS AND DISCUSSION

As already reported in previous studies on MISFET structures
based on p-type organic semiconductors[4-7], majority carriers
are injected into the organic semiconductor through the ohmic
source and drain gold contacts. The current arises then from
the formation of an accumulation layer, as confirmed by capaci-
tance-voltage measurements on MIS diodes[7]. The active structure
involves thus a thin channel at the interface between the organic semiconductor and the insulator, this channel lying in parallel with the bulk of the semiconductor. The observed total current represents then the sum of the drain current, I_D, and of
the ohmic current in the bulk, I_Ω. At saturation, they follow
the equations (1), (2) and (3) :

$$I_{Tot.} = I_{D,Sat.} + I_\Omega \qquad (1)$$

$$I_{D,Sat.} = \mu C_{ox} Z / 2L (V_G - V_0)^2 \qquad (2)$$

$$I_\Omega = g_\Omega V_D = (nq\mu Zh / L) V_D \qquad (3)$$

where C_{ox} is the capacitance of the oxide layer, Z and L the
channel width and length, h the thickness of the semiconductor
layer, n the density of free carriers, and V_G and V_D the gate
and drain voltage.

When an as deposited α-6T film is used as active layer, the
obtained amplification characteristics, Fig. 2, show that even
at V_G = 0 V, a high current is flowing through the device. This
can be attributed to an ohmic current, underlining the large
contribution of the bulk conductivity of the semiconductor,
which, even of the order of 3.7×10^{-7} Scm^{-1}, still leads to
a noticeable effect on the observed total current.

Conditions for improving the characteristics of an organic FET
can be deduced from the equations (1) (2) and (3). As a matter
of fact its optimization will involve at first the increase of
the ratio I_D/I_Ω. At saturation, this ratio value is given by
equation (4) :

$$(I_D / I_\Omega) = C_{ox} V_D / 2 \, nqh \qquad (4)$$

The increase of this ratio will be obtained by decreasing the
free carrier density n and the thickness h. As the channel
current should be as high as possible, it appears not well-
advised to decrease the free carrier density. The most efficient
way to improve this FET appears thus to decrease its thickness.
As a first consequence, an accumulation layer MISFET must be
a thin film device, with thicknesses lower than 100 nm. These
devices are thus better described under the term of Thin Film
Transistors. In order to confirm this conclusion, we performed
the study of the effect of conjugated oligomer thickness on
the transistor characteristics. We first checked that the
electric parameters are not altered by decreasing the film
thickness. As a matter of fact, it has been already reported in
the literature that the carrier mobility in casted films of
polyhexylthiophene drops dramatically when the film thickness
is lowered.[5] In the present case of α-6T, we observed that
when going from 200 to 10 nm for the layer thickness, the mobility value decreased only by a factor of 4, which underlines
the great homogeneity of these vacuum evaporated films.

The variation of the total source-drain current as a function
of the gate voltage at a drain bias of -10 V is given, in semi-
log scale, in Fig. 4. for two MISFETs with two different thicknesses : a 160 and b 14 nm. The results illustrate the large
improvement brought by reducing the semiconductor layer thickness. At zero gate bias, the observed current equals the ohmic
current. When decreasing the thickness, this current decreases
by the same amount, whereas the decrease in $I_{D,Sat.}$ is only due
to the lowering of the carrier mobility, as given in Eq. (2).
At zero gate voltage V_G, I_D = 0.14 nA for sample a, and 0.02 nA
for sample b, a decrease comparable to that of the thickness.

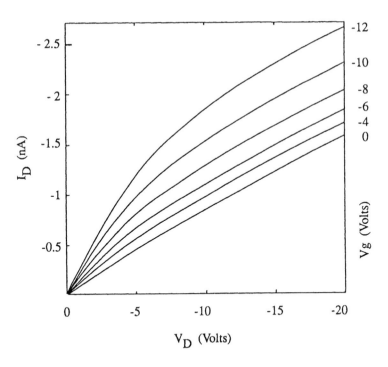

Fig. 2. Source-drain current (I_D) versus source-drain voltage (V_D) of α-6T based FET at various gate voltages (V_G). The thickness of the as-deposited α-6T layer is 160 nm.

In order to improve the device characteristics, the annealing of the device has been performed by heating it at 120 °C during 3 hours in ambient atmosphere. The I_D-V_D characteristics of the device obtained after heat treatment are drastically changed, as shown in Fig. 3. A simultaneous decrease of the ohmic current and an increase of the channel current are observed, leading to a significative improvement of the device characteristics. The field-effect mobility μ, which can be determined from the equation (2) by the $(I_{D,Sat.})^{1/2}$ versus V_G plot, rises to the value of 10^{-3} cm^{2}V^{-1}s^{-1}. To our knowledge, this is the highest reported value for a field-effect mobility in an organic semiconductor. It is 100 times higher than the one obtained on poly methylthiophene[4], and 10 times higher than the mobilities measured on poly hexylthiophene[5] and on polyacetylene[7]. We have already reported a value of about 10^{-2} cm^{2}V^{-1}s^{-1} for the carrier mobility of α-6T, based on Schottky diodes characteristics.

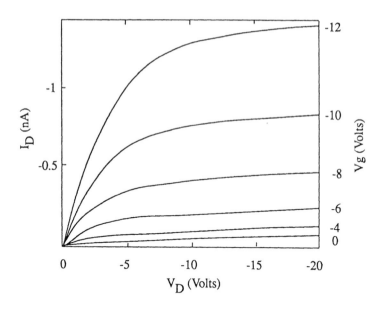

Fig. 3. Source-drain current (I_D) versus source-drain voltage
(V_D) of the same α-6T based FET as in Fig. 2., after a heat
treatment of 3 h in air at 120 °C.

This discrepancy in mobility values has often been observed in
the literature when two very different techniques are used for
measuring the same parameter. Nevertheless, the results brought
here give a confirmation of the very high mobility values wich
can be reached by using conjugated oligomers instead of their
corresponding polymers.

It must be reminded that before heat treatment, the mobility
value derived from the results given in Fig. 2. lead to a value
of about $3 \times 10^{-4} cm^2 V^{-1} s^{-1}$. Due to the increase of mobility after
annealing, the lowering of the ohmic current observed in Fig. 3.
must be ascribed to an undoping of the α-6T. By the use of IR
and visible absorption spectroscopy, together with elemental
analysis, we confirmed that the conjugated oligomer α-6T is
chemically stable up to 150 °C, ruling thus out any chemical
modfication of the compound during heat treatment. The mobility
increase can then be ascribed to some physical change, e.g. to
crystallization or conformation change in the molecule.

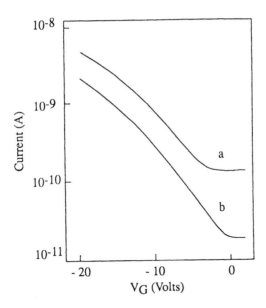

Fig. 4. Source-drain current versus gate voltage at V_D = -10 V
for two annealed α-6T MISFETs with two different thicknesses :
a 160 nm, and b 14 nm.

When the gate voltage V_G is increased to -20 V, the observed
total current becomes 4.7 nA for sample a, and 2.2 nA for sample
b. The respective values of the ratio I_D / I_Ω are thus 110
and 34 respectively for samples b and a, which represent an
increase of a factor of 3.2 for the thinner device. These
values also represent the I_{on} / I_{off} ratio, underlining the
great advantage brought by the use of a thin film device.

4.   CONCLUSION

Unlike conventional devices, organic-based MISFETs operate
through the formation of an accumulation layer channel. In
the model presented here for describing these organic devices,
we propose to take account for the ohmic current in parallel
with the channel current, originating from the non-rectifying
character of the source and drain contacts. It is shown that
the improvement of organic MISFETs can be achieved by increasing
the carrier mobility of the organic semiconductor, and also
by increasing the channel-to-ohmic current ratio. The first

first point has been reached by annealing the α-conjugated sexi-
thienylene, leading to a field-effect mobility of about 10^{-3} cm^2
$V^{-1}s^{-1}$, the highest value up to now reported for an organic semiconductor. The second point has been achieved by reducing the
film thickness of the semiconducting layer. These thin film
transistors possess very promising characteristics.

REFERENCES

(1)　Weinberg B R, Gau S C and Kyss K, 1981 Appl. Phys. Lett.
　　　38 555

(2)　Fedorko P and Kanicki J 1984 Thin Solid Films 113 1

(3)　Glenis S, Horowitz G, Tourillon G and Garnier F 1984 Thin
　　　Solid Films 111 93

(4)　Tsumura A, Koezuka H and Ando T 1988 Synth. Metals 25 11

(5)　Tsunoda S, Koezuka H, Kurata T, Yanaura S and Ando T 1988
　　　J. Polymer Sci. B, Polymer Phys. 26 1697

(6)　Assadi A, Svensson C Willander M and Inganäs O 1988 Appl.
　　　Phys. Lett. 53 195

(7)　Burroughes J H, Jones C A and Friend R H 1988 Nature 335
　　　137

(8)　Fichou D, Horowitz G and Garnier F J.Chem.Soc. Chem. Comm.
　　　submitted

Correlations between active agents and electrically conducting polymers

Herbert Naarmann

1. INDRODUCTION

"Look to the past to understand the present and learn for the future".
This maxim applies not only to current affairs but also to scientific and technical developments, such as those in the field of
electrically conducting organic materials.
Ever since this fascinating field was discovered more than 20 years
ago [1], it has been the subject of thorough research. It took a long
time to learn that the benefits of electrically conducting polymers,
which can be reproducibly manufactured as powders, films (also of the
oriented type), coatings, fibres, mixtures, etc., lie less in providing
substitutes for conventional metals than in opening up new areas of
application. This calls for creativity and innovation.

2. POLYPYRROLE PROPERTIES

Polypyrroles have been taken as examples of polymers that contain
heteroatoms in order to demonstrate how **the properties of materials can
altered:**
* By varying the counterions
  This allows controlled release of built-in counterions, e.g.
  dyes, drugs.
  Uses: Sensors, surgical plasters containing active ingredients,
        membranes [2]
* By varying the polymerization conditions [3] to influence
  thickness of films, conductivity, conductivity gradient, film
  elasticity
  Uses: Electrodes, rechargeable batteries, protection against total
        discharge, ELMI shielding

* By surface deposition for antistatic finishing, optical storage
  systems [4], manufacture of printed circuit boards [5],
  piezoceramics [6].

These examples serve to illustrate the wide range of potential
applications.

2.1 The influence of the counterions

The conductive salt (x⁻) used in the electrochemical synthesis is
contained in the polypyrrole film as a rechargeable counterion [3].
Interesting conductive salts that affect optical activity in polypyrrole
(+) or (-)-camphor sulphonic acid, used for racemate separations,
heparin [7] or other drugs such as Monobactam [8]. The process of
charging and discharging is illustrated in Figure 1.

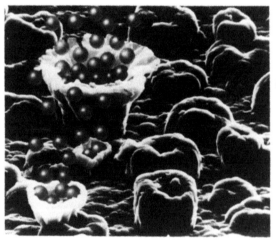

$\left(\text{X}^{-}\right)$: Indigocarmine

Lucifer yellow CH

Cu-phthalocyanine

Monobactam-aztreonam
(Penicillin)

Active ingredient

Incorporation   and   Release

$$\text{X}^{-} \downarrow \qquad\qquad \uparrow \text{X}^{-}$$

Polypyrrole film

o charging by a definite current, counter ion e.g. Heparin

o discharging and releasing of the counterion (-the amount of released
  active ingredient is equivalent to the current-) into the counter
  electrode, e.g. skin, membrane, substrate

Fig. 1.   Charging and discharging process of polypyrrole

2.2   Polymerization conditions

Figures 2 and 3 show continuous processes. They were developed from batch processes and use a rotating drum (Figure 2) or moving belt (Figure 3) as anode. In both cases, the counterelectrode is equidistant. The factors that affect the continuous production of homogeneous polypyrrole films are the residence time at the anode or the speed of rotation, the current density of the monomers, the concentration of the conductive salt, etc. In practice, the process consists of withdrawing a polymer film directly from the electrolyte containing the pyrrole and the salt and winding it up [9].

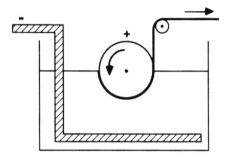

Fig. 2.   Continuous production of polypyrrole. The anode was in drum form and allowed the polymer film to be withdrawn and wound up direct from the electrolyte solution.

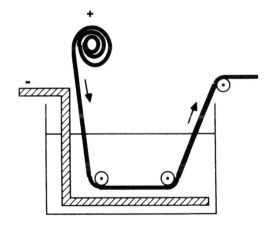

Fig. 3.   Continuous belt process. The anode is a belt on one side of which the polypyrrole is deposited. If two counterelectrodes are used, the polypyrrole can be deposited on both sides.

Depending on the reaction conditions, it is possible to produce fle-xible films with gauges of 30 m to 150 m that can be readily wound. The belt process (cf. Figure 3) is mainly resorted to if conductive materials, e.g. polymers extended with carbon black, metal foil, or fabrics produced from carbon fibres, are to be laminated with the polypyrrole or have to be electrochemically modified by other means. No problems are involved in laminating these to one or both sides of the polypyrrole film.

The main properties of polypyrrole are listed in Table 1.

Conductivity (S/cm)	10^{-4} to 10^{+2}
Film gauge a) Self-supporting films	ca. 30 upwards
(μm) b) Coatings	0.01 upwards
Specific surface	
Nitrogen surface (m^2/g)	5 to 50
X-ray crystallinity (%)	15
Oxygen absorption % wt/day	0.1/30

3.   APPLICATIONS

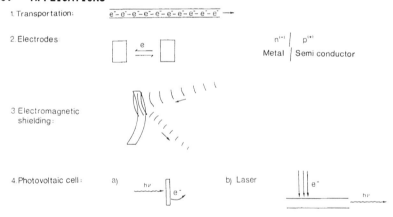

1. Transportation: $\overline{e^- \cdot e^- \cdot e^- \cdot e^- \cdot e^- \cdot e^- \cdot e^- \cdot e^- \cdot e^- \cdot e^-}$ ⟶

2. Electrodes

3. Electromagnetic shielding:

4. Photovoltaic cell: a)   b) Laser

5. Antistatic equipment, immersion heating systems, sensors, transistors or reversible batteries, electronic devices

Fig. 4.   Some typical applications for conductive polymers (e. g. polypyrroles)

Since polypyrrole film has even higher conductivity than plastics extended with conductive fillers, it is only logical to use it for electromagnetic interference shields. The results presented in Figure 9

demonstrate the good shielding effect of polypyrrole film over a wide
range of frequencies in a radiation field. The attenuation required in
this case is about 40 dB, i.e. the amplitude of the electromagnetic
radiation should be reduced by a factor of about 100.

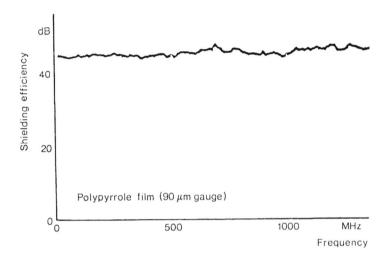

Fig. 5. Shielding efficiency as a function of frequency
         (ASTM ES 7-83)

3.1 Applications in which the electrochemical reversibility is exploited

Polypyrrole is a suitable electrode material for rechargeable electrochemical cells. The advantage of polymer electrodes is that they can be
easily shaped. As a result, they allow new battery types to be designed,
e.g. for the electronics sector, and new production methods involving
lower costs to be adopted.

Cells with polypyrrole and lithium electrodes have been developed
jointly with VARTA Batterie AG, Kelkheim. Their energy per unit mass and
their discharge characteristics are very much the same as those of the
nickel/cadmium cells now on the market. More than 500 charging and
recharging cycles were achieved with laboratory cells.

Two types of cell in which polypyrrole has been used are shown in Figure
6. In the flat cell, the polypyrrole and lithium films are sandwiched;
in the cylindrical cells, the two films are wound concentrically. These
cells give an idea of the wide range of possibilities offered by polymer
electrodes in the form of films [10].

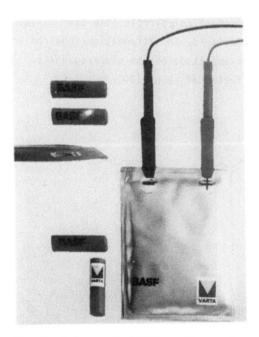

Fig. 6   Polypyrrole/lithium flat and round cells

3.2   Chemical modification of surfaces [11]

An interesting process variant is chemical modification with hetero-
cycles, e.g. thiophene or pyrrole. By this means, conductive coatings of
e.g. polypyrrole, can be produced on Ultrason, Teflon, Styrolux, PVC, or
other polymer films, which (to an extent depending on the thickness of
the polypyrrole coating) may be transparent and antistatic with
conductivities of about 10^{-3} S/cm (Figure 7).
Ceramics and glass can also be chemically modified (Figure 8).
Porous materials such as wood, fabrics and open-celled foams, e.g.
Basotect, can also be modified and made antistatic by this method
(Figures 9, 10).
Conductive powders, e.g. of polypyrrole, with particles of about 0.1 μm
diameter and conductivities of up to 10 S/cm can also be produced by
chemical oxidation and can be incorporated as fillers in thermoplastics.

Fig. 7 View through an Ultrason film made antistatic and conductive by chemical modification with polypyrrole

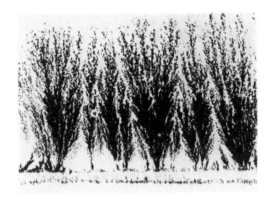

Fig. 8 Growth of polypyrrole dendrites in a porous ceramic

Fig. 9 Beech coated with polypyrrole; conductivity 10^{-3} S/cm

Fig. 10   Nylon fabric coated with polypyrrole; conductivity 10^{-3} S/cm

4.   POLYENES - SENSORS AND SIGNALS

Polyacetylenes are ideal model compounds for measuring such physical parameters as electrical conductivity, magnetic susceptibility, carrier mobility etc. in the study of chemical reaction mechanism.

One problem is the controlled preparation of soluble systems, which should be catalyst-free. It is rather difficult to prepare metal-free polyenes by the Grubbs-Schrock [12] or Feast methods [13].
An other possibility is the use of solubilizing side groups [14] to prepare backbone polyenes e.g:

Type I $\left(\begin{array}{c}\diagup C\diagdown_C\diagup C\diagdown_C\end{array}\right)\begin{array}{c}R\\ |\\ C\\ |\\ R\end{array}$ R = CN , CONH$_2$, phenyl

4.1   Synthetic methods

With the Wittig reaction it is easy to synthesize polymers with polyene side groups [15].

Type II $(-CH_2-CH-)_n$

CH$_2$O—polyene

Polyene: e. g.

Retinyl

Type III $(-CH_2-CH-)_n$ or higher molecular polyene side groups

$CH=CH-$polyene

In all cases the basic idea was to use compounds types I - III as molecular wires to transport electrical charges and to conduct impulses.

A completely new class of sensors are those based on adrenaline (or homologous derivatives), Type IV [16].

Replacing the methyl group in adrenaline with a polyene side group in the presence of mild oxidizing agents e.g. $K_4(Fe\ CN)_6$ creates an impulse - as in the formation of the quinoid cyclic system by ring closure to give an indol derivative (pyrrole). In such a model system the quinoid system acts as an intramolecular doping system, forming an intramolecular CT complex that is electrically conducting, the π-double bonds in the polyene allowing the electrical charge transportation. (In the body, the adrenaline is responsible for nerve stimulation).

Type IV

This substance (Type IV) can be used as a powerful colorimetric sensor of changing electrode potentials.

4.2   Variations of structure isomers

Tab. 2: Correlation of structure and properties

		x) Ep/V	polymers	σ S/cm
	pyrrole	1.2	+	$10^{-4} \rightarrow 10^{+2}$
	isoindole	1.1	+	$10^{-4} \rightarrow 10^{0}$
	indole	1.45	- (dimers)	10^{-6}
	indolicine	1.2	+	$10^{-6} \rightarrow 10^{0}$

x) cyclic voltammetric data in 0.1 M tetraethylammonium fluoroborate in CH_3CN versus SSCE.

As shown in Table 2, there is a strong correlation between structure and properties but the main interesting thing was the idea to substitute the listed monomers e.g. by $-SO_3^{(-)}$ and to have a monomer which acts also as a counterion. Due to this dual function, these materials are of interest for electrodes and battery applications [17].

5. POLYTHIOPHENES

Due to the preparation technique with boronic acid [18]

etc. oligomers

it was possible to prepare well-characterized oligothiophenes [19].
A direct route to isothionaphthene is described in [20].

Table 3 and Figure 11 show the correlation between structure and properties. The usefulness of bithiophene or higher oligomers lies in the
fact that they can be easily oxidized by air, forming excellent depo-
sits (thickness = $<1 \mu m$), on natural or synthetic material with sufficient conductivity (10^{-2} S/cm) to work as antielectrostatic coatings
e.g. for electronic devices [21].

Table 3:

n	n 1	2	3	4	5	6	7	8
mp (°C)	-30	33	95	214	258	307	324	339
E_{nx} (Volts)[x)]	0.78	0.43	0.21	0.12	0.08	0.04	0.03	0.025
UV visible (nm) ($CHCl_3$)	280	300	354	390	416	434	442	449
Mass spectrum (m/e)	84	166	248	330	412	494	576	658

[x)] Ag/Ag^+ ; 0.5 M $LiCl_4$/ Propylenecarbonate

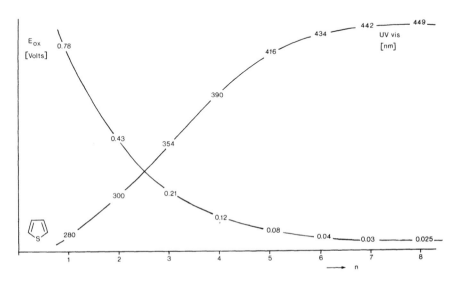

Fig. 11. UV-absorption and oxidation potentional as a function of the
chain lenghth (n)

6.   REFERENCES:

1   **DE-Pat.** 1178529   11.04.63/11.08.65 BASF AG/FRG, Naarmann H, Beck F
    and Kastning E G
    GdCH-München, Halbleitertagung 12. Okt. 1964 München, Naarmann H,
    Beck F
    Naarmann H 1969 **Angew. Chemie Internat. Edit. 8** 915

2   **US-Pat.** 4585536 23.04.83/31.10.84 BASF AG/FRG

3   Naarmann H 1987 **Macromol. Chem. Macromol. Symp. 8** 1

4   **DE-OS** 3601067   16.01.86/29.07.87 BASF AG/FRG

5   **DE-OS** 3535709   05.10.85/09.04.87 BASF AG/FRG
    DE-OS 3520980   12.06.85/11.06.86 BASF AG/FRG

6   Schnöller M, Wersing W and Naarmann H 1987 **Macromol. Chem.
    Macromol. Symp. 8** 83

7   **DE-OS** 3607302   06.03.86/10.09.87 BASF AG/FRG

8   **DE-OS** 3630707   10.09.86/17.03.88 BASF AG/FRG

9   **US-Pat.** 4468291 14.07.82/28.08.84 BASF AG/FRG

10 Bittihn R 1989 **Kunststoffe 79/6** 530

11 **BASF Research and Development** KVX 8611 d, Oct. 1986 pp 37 - 40

12 Swager T M, Dougherty D A and Grubbs R H 1989 **J.Am. Chem. Soc.**
    111 4413

13 Edwards H H and Feast W J 1980 **Polymer 21** 595

14 **DE-OS** 3114342   09.04.81/04.11.82 BASF AG/FRG

15 **DE-OS** 3701007   15.01.87/28.07.88 BASF AG/FRG

16 **DE-P** 3942282.8   21.12.89 BASF-AG/FRG

17 **DE-OS** 3425511   17.07.84/16.01.1986 BASF AG/FRG

18 Miyaura N, Yanagi T and Suzuki A 1982 **Synth. Commun. 11/7** 513
    Roncali I, Garnier F, Lemaire M and Garreau R 1986 **Synth. Met.**
    15 323

19 Martinez F, Voelkel R, Naegele D and Naarmann H 1989 **Mol. Cryst.
    Liq. Cryst. 167** 227

20 **DE-OS** 3508209   08.03.85/11.09.86 BASF AG/FRG

21 **DE-OS** 3502937   30.01.85/31.07.86 BASF-AG/FRG

Electronic conducting polymers: New results on low, medium and high technology applications

Eugène M. Geniès

Electrochimie Moléculaire, Laboratoires de Chimie,
Département de Recherche Fondamentale,
Centre d'Etudes Nucléaires, 85x, 38041 Grenoble, France

ABSTRACT: This chapter is a review of some selected papers which are used to discuss of the problems concerning the studies of electronic conducting polymers on low, medium and high technology applications. What we call low technology is typically concerned with products which can be made in form of a blend of electronic conducting polymers with conventional polymers for applications which need a large quantity of material such as for ESD and EMI shielding. With medium technologies, there will be specially some applications which need specific counter ions, as for membranes. High technology is concerned with products which need a complex processing, such as batteries, displays, smart windows, sensors, electronic components, etc...

1. INTRODUCTION

In the last few years progress in the processing of electronic conducting polymers has been considerable. Interest is primarily for the numerous applications of these materials and also for fundamental aspects. Effectively, new processing can give new information on: structures of polymers, better defined materials, more conductive and more cristalline materials, etc...The aim of this chapter is not to present an exhaustive review, but to focus on some selected papers and results obtained in Grenoble in order to discuss the approach to the electronic conducting polymers when we want to use it for different applications in low, medium and high technology areas. What we call low technology is that which is concerned with products, such as blends containing some amount of conducting polymers. This is very important with new methods of processing. The applications are, of course, primarily in the area of the antistatic (ESD) and EMI shieldings. Material composites and blends can also be used for applications of medium technologies as support for chromatography and other chemical analysis, for catalysis, for different kinds of membranes, etc... Included in high technology applications are all the products which need a complexe processing. Fot example: displays or smart windows, batteries, electronic components, photoelectrochemical or photoelectrical devices, electroluminescence devices, sensors, etc...

2. LOW TECHNOLOGY APPLICATIONS

These applications are dominated by the current progress in new methods of processing. It will be discussed in this conference by J.-E. Österholm[1].
The ideas and concepts which are used to improve the processing of electronic conducting polymers are the following: chemical preparations of molecular blends, of soluble and/or fusible polymers, compatible polymers, synthesis of substituted polymers which bear the preceeding properties, synthesis of soluble precursors, synthesis of stable oligomers, chemical preparations in solution or gas phase, preparation by sublimation of polymers or oligomers, plasma methods, Langmuir-Blodgett techniques and thermic treatments.
It is relatively easy to chemically prepare composites and blends with polyaromatics, i.e. polypyrrole, polythiophene, polyaniline, etc... It needs a solution of the monomer, an oxidant such as iron chloride and other components, creating optimal mecanical properties and the ability to process. Others components can also have an active role when they bear specific properties[2]. The doping anion then confers a certain function/property upon the polymer[3]. The use of such conducting polymers has been described for : Electrochromism[4], photoelectric conversion, electroluminescence, highly dispersed noble metals[5], catalysts[6], charge controlled transport membranes[7], sensors[8], controlled release membranes[9], biocompatible materials[1], and detection of metal traces[10].
An important area of development is concerned with blends and composites. For most of applications like ESD or EMI shielding, it is an advantage, with regard to the price and processability, to prepare blends with smaller amounts of conducting polymers (1 to 10%). For polyaromatic conducting

polymers polyaromatics, it is easy to create blends by preparation in presence of a solution of the conventionnal polymer. Of course when the polyaromatic can be prepared in water as in the case of polypyrrole, and when the conventional polymer can be obtained at least as a very thin water dispersion it is possible to have blends with enough good electric properties for antistatic applications on an industrial scale. It is possible to obtain conducting material in bulk, on surface or with a concentration profile. Some possibilities for copolymerisation exist, for instance with pyrrole and naphtol which was achieved by electrochemistry[11], but which may also prepared by chemistry. Also polypyrrole-PMMA films can be prepared from layers of a solution containing, pyrrole, PMMA an benzoic acid in 2-butanone, in an aqueous solution of potassium persulfate. The films are 0.1 to 10 μm thick with a concentration profile of PMMA and polypyrrole. The conductivity is from 10^{-8} to 10^{-1} S/cm[12].

It is also relatively easy to prepare composites with natural polymers. For example, the synthesis of polyaniline / polysaccharides from dialysis membranes or dinitrocellulose films which are first dipped in a solution of aniline chlorhydrate 1M and then for one hour in persulfate solution at 4°C. The composite keeps the mecanical properties of the polysaccharide matrix and remains perfectly stable over time[13]. The literature is very rich in this area and corresponds probably to more than one hundred references.

Simple chemical methods using solutions facilitates coating of conducting polymers on any material, conductive or not. For instance, Milliken and our group[14,15] have made conductive textiles with excellent mecanical properties and a resistivity of about few tens ohms.

After coating , on any insulating materials of polypyrrole (by dipping in the proper solution), it is possible to electrochemicaly deposit a metal such as copper[16]. Such process can be used as improvement of galvanoplasty, for electrical circuits or for excellent EMI shielding. After that there is little concern whether the polypyrrole will degrade or not. Such process could be usefull also to prepare catalyst for chemical industries by coating a powder in a fluidized bed.

An important field is concerned with fusible and soluble conducting polymers. In this area, the most important progress has been achieved by the members of "Nordisk Industrifond Research Project"[1] with poly(3-alkyl-thiophenes).

However, for soluble polymers in conductive form and for fusible conducting polymers the results are still limited. For solutions the question it is still under investigation, to determine whether they are true solutions or dispersions. Polyaniline is possibly the polymer which allow easy preparation of solutions[17]. An interesting possibility for any polyaromatics is to use in the preparation a large counter anion which will confer properties of solubility. For instance polyaniline has been prepared in presence of a derivative of chitine (CHITOSAN)[18]. The resulting product is soluble in 2% acetic acid solution and gives very strong films with 10^{-2} S/cm conductivity. We prepare have prepared a polypyrrole with specific counter ions which gives apparently a fusible polypyrrole[19]. Such kind of materials can be used to prepare extrusion or injection products for large applications.

A particular mention must be made of the colloïd solutions or dispersions which are obtained in the presence of submicronic particules such as latex[20]. Authors consider that polypyrrole is coated around the micromarticules or that it may act a stabilisor which is adsorbed on the polypyrrole and limits its growth[21,22].

Soluble polymers, even in water, have been described for substituted polythiophene and polypyrrole, with a sulfonated alkyl chain[23,24]. In the area of soluble polymers it is necessary to mention the techniques to prepare a soluble precursor which is activated to a vinyl polymer by heating to a temperature of ca 100°C[25,26].

For low technology applications, it must be recalled, that for any blends or composite preparations, the conducting polymer must be compatible from both the chemical and biochemical point of view. This can be achieved with the counter ions but of course also by modifications of the polymer chain[27].

The preparation of stable and pure oligomers are of large interest for any level of technology[28].

3. MEDIUM TECHNOLOGY APPLICATIONS

In this domain the process of preparation of the material containing the conducting polymers has already been discussed. It is mainly concerned with the use of specific anions or cations. Lets mention : Material for controlled high-performance liquid chromatography[29]. Membranes with specific ionic conductivities[30]. Preparation of membranes with Nafion and conducting polymers which can be use for electrolysis or demineralisation of water[31],[32]. In the case of material for catalysis, the possibilities are considerable. Unfortunatly the authors present a study made by electrochemistry. It should be very important in most cases that they extend the process of preparation to a chemical one in order to be able to prepare large amounts of material for industry. There are many such catalyst species: metal particules[33], clay[34], zeolites[35], specific anions[6], etc...

For the micro-wave applications many results have been obtained with polyaniline[36]. However, it is not possible to mention everything. Possible of great use also is the production of supercapacitors, specialy because we know now how to coat aluminium with polypyrrole[37].

4. HIGH TECHNOLOGY APPLICATIONS

Concerning sensors, although the principes are simple the technology is not, because it is generally encapsulated in an integrated circuit. This is found in sensors for gas[38], or for analytical chemistry studies of solution[3]. Many specific sensors could be made by the immobilization of specific ligands of metallic cations within the conducting polymers[3]. Some bio-sensors have already been manufactured for glucose and urea[39]. They work by amperometric measurement, because of the immobilization of the specific enzyme. They can also work by potential measurement[38] of the immobilised bio-compound concerned by the determination. In principle, sensors could be prepared with various antibodies, for example in order to detect if a person has been in contact of a given virus.

PHOTOELECTROCHEMICAL AND PHOTOELECTRICAL DEVICES

For applications in non-linear optics, J.-L. Bredas[40] must be refered, to D.T. Bloor[41] for integrated optics, to R. Friend[42] for semiconductors devices, and to S. Roth[43], for molecular electronic, conductivity and photoconductivity, . We have observed very interesting photo-electrical properties of polyaniline films and other polyaromatics with respect to the electrolyte, concentration and wavelenght of light[44]. The direction and level of the current photoresponse is function of the relative number of transport carriers in the polymer and in the solution.

In principe these properties could be applied to make a cheaper video camera. It is possible to imagine using layers of electonic conducting polymer (and proper additives), electroluminesent windows or screens. This could be obtained by two redox systems, one reduced and one oxidized which recombine causing light emmission.

Electronic conducting polymers such as polypyrrole, polythiophène, polyaniline and other polyaromatics can be grown from a monomer solution and the light irradiations of semiconductors as silicon, CdS, etc...[45] or by the excitation of a small layer or amount of the polymers which also have rather good semiconductor properties[46]. This has been done with electrodes, but it can propably be carried out on a powder of the inorganic semiconductor or a powder of the polymer itself. If the solution also contains specific anions for specific properties, for instance catalytic, this allows the possibility for the prepararation of large amounts of useful materials.

BATTERIES

Almost all conducting polymers have been investigated for battery applications in organic media as well as in aqueous media. The following table lists some of these :

Positive electrode .	Negative electrode .	Electrolyte Solvant, salt	References
PPP	Pb	H_2O, H_2SO_4	47
CHx	CHx	PC, $LiClO_4$	48
CHx	Li	PC, $LiClO^4$	49
PANi	Li/Al	PC, $LiClO^4$	50
PANi	Zn	H_2O, $ZnCl_2$	51
PPy	Li/Al	PC, $LiClO_4$	52
Pth	Li/Al	PC, $LiClO_4$	53

PPP=poly(p-phenylene),   CHx=polyacetylène,   PANi=polyaniline,   PPy =polypyrrole, Pth=polythiophène, PC=propylene carbonate.

Such studies were carried out not only by academic laboratories but also by many important industrial companies. However, only Bridgestone[54] have marketed a rechargeable button-cell with electrochemical polyaniline as the positive electrode.

The market for secondary batteries is very large and requires elements with completly differents characteristics, to start a car, for electric transportation systems, for back-up memory in computers, for energy storage, for emergency light, etc... A polymer cell can have a place in such market. However one important difficulty here is the stability of the polymer upon cycling. In our studies it was observed[50], that polyaniline was the best from this point of view as to its behaviour in floating. In addition to instantaneous power, the capacity in Ah/kg is another characteristic of conducting polymer materials which is rather too low. For polyaniline the value for the polymer and the salt is about 120 Ah/kg[55]. This value is higher for polyaniline than other conducting polymers, because in the redox process, polyaniline loses protons and then the size of the charge interactions decreases and the polymer can be more oxidized. However, this value is still too low for transportation systems. For pure polymer, even with an hypothetical new molecule it is really impossible to obtain a better value because of the large electrostatic interactions[56] in the polymer which theoritically limits the number of positive charges in the material. However, we can imagine a conducting polymer in which the doping agent also bears a redox behaviour.

It is well known, that a better system in terms of capacity is the sodium-sulfur battery. A good test could be for a conducting polymer containing sulfur redox systems. A first approach to this was investigated by M. Armand[57] using some carbon species with sulfur, and by A. Lemehauté[58] using PVC (precursor of CHx). In the latter, for the first cyle the author obtained 1000 Ah/kg. Unfortunatlly it seems that the cycling behaviour was very bad. We also briefly investigated the behaviour of polyaniline and sulfur[59]. The best recent results were from L. de Jonghe and S. Visco at the Lawrence Berkeley laboratory[60]. The positive active material is made by a polymer of thiadiazole rings with sulfur-sulfur bonds in the oxidation state, and it depolymerizes to sulfur-lithium bonds in the reduced state. The authors claim 160 to 180 Watts per kilogram of batteries. These values are good enough for electric vehicles. However, Berkeley's system is not yet the best, and we must synthesize a modified conducting polymer which can bear the sulfur-sulfur redox reaction. Hopefully this will be done soon and many electric vehicles will travel in the next 5 years in our large cities with the assistance of an electrical conducting polymer.

ELECTROCHROMIC DEVICES

The use of electronic conducting polymers is envisaged to be for electrochromic devices (ECDs), especially for the construction of: displays, active optical filters, automotive rear-view mirrors, and "smart windows" with adjustable absorption and reflection in the visible and near infrared wavelength ranges in oedre to control the energy transmission of buildings[61].

The most widely applied electrochromic substance is tungsten trioxide (WO_3). But this material must be set in opposition to another electrode. Effectively, ECDs are similar to an electrochemical cell or battery. They consist of two electrodes and an electrolyte. At least one of the two electrodes must be transparent and one must be electrochromic. Both electrodes must be able to accumulate the same amount of charge. When one electrode is in the oxidized state, the other must be in the reduced state, and each of these states must be compatible, hence in the more coloured or in the more transparent state. Combinations between electronic conducting polymers and inorganic materials include the following[62]: Polyaniline (PANI) (dark blue in oxidised form, light yellow in reduced form) with Polythiophenes (Pth) (dark red in reduced form and light blue in oxidized form); PANi with WO_3, and IrO_2 with Pth. WO_3 is the best transition metal oxide exhibiting cathodic coloration and IrO_2 is the best investigated anodically coloring material.

Many other electrochromic systems have been studied. Two well known systems are ferric ferrocyanide (prussian blue)[63] for inorganic compounds and viologene for organic ones, which have both been associated with conducting polymers[61].

Polyaniline has been largely investigated as an electrochromic material[64]. It has also been combined with IrO_2, to increase the coloration efficiency and the percolation of electrons to IrO_2[65]. The advantages of growing electronic conducting polymers in the presence of electrochromic materials, (neutral or anionic) are considerable and have not yet been fully studied. Many others possibilities have still to be investigated[2].

The advantages of organic electrochromic materials, including polymers, are their relatively rapid switching capability and their simple handling for construction of ECDs. However, undesired and irreversible side reactions are a general disadvantage of organic materials and are a reason for their rather short lifetime and poor UV stability. It seems that the stability of electronic conducting polymers to the light must be largely improved by adapting an electronic circuit in order to collect the hole-electron pairs which are created in the conducting polymer.

The choice of an appropriate electrolyte has a strong influence on the properties of the electrode material.

For instance, IrO_2, presents a faradic behaviour with coloration switch in aqueous medium and only a capacitive behaviour in organic medium without switching of coloration[66]. Polypyrrole-latex presents in organic and aqueous medium only capacitive behaviour with almost constant optical properties. This seems to be the case for many electronic conducting polymers which have been prepared in presence of very large anions. It became very difficult to undope it by electrochemical reduction, it remains conductive in a larger range of potential with accumulation of charge as a capacitor. This property has not been exploited for all conducting polymers. It can be useful for supercapacitors and also in ECDs as an electrode which acumulates charges with a constant optical properties.

Liquid and polymeric electrolytes were employed mainly during the period of development of electrochromic systems. Currently vapor deposited oxides as ion conductors are being developed, e.g. ZrO_2, HfO_2, $TaO2$, SrO_2 and fluorides, e.g. CaF_2, MgF_2, and CeF_3[67].

5. CONCLUSION

There are in the literature a considerable number of patents. The question can be asked, if all these patents are valuable. It is probably not the case. For instance, with polyaniline batteries, there are several tens of patents. Polyaniline has been known for about 150 years and Buvet et al. presented a patent of a battery in aqeous medium 25 years ago. However, knowledge in laboratories in order to make real applications is important and this can be exploited and valued. The knowledege stemming from academic laboratories is well known. Unfortunately, industrial laboratories have accumulated a considerable amount of very interesting results of which the scientific community are not informed. If large companies do not whish to exploit such results, it would be good to propose it to smaller companies and/or to publish it.

We can soon expect some new interesting basic molecules to appear. We can also expect that in the area of electonic conducting polymers, delays between fundamental results, applied results, developments and the marketing of a product are shorter than in other material sciences.

References

1- J.-E. Österholm, this conference.

2- H. Naarmann, this conference.

3- T. Shimidzu, Reactive Polymers, § (1987) 221.

4- M. Salmon, A.F. Diaz, A.J. Logan, M. Knounbi and J. Bargon, Mol. Cryst. Liq. Cryst., 83 (1982) 265.

5- P. Ocon Esteran, J.-M. Leger, C. Lamy and E.M. Genies, J. Appl. Electrochem., 260 (1989) 462.

6- G. Bidan, E.M. Genies and M. Lapkowski, Synth. Metals, 31 (1989) 327.

7- T. Shimidzu, O. Ohtani, T. Iyoda and K. Honda, J. Electroanal. Chem., 224 (1987) 123.

8- A. Boyle, E.M. Genies and M. Fouletier, J. Electroanal. Chem., 279 (1990) 179.

9- Q.-X. Zhou, L.L. Miller, and J.R. Valentine, J. Electroanal. Chem., 261 (1989) 147.

10- T. Iyoda, A. Ohtani, T. Shimidzu and K. Honda, Synth. Met., 18 (1987) 725 and reference therein.

11- B.D. Malhotra, N. Kumar, S. Ghosh, K.K. Singh and S. Chandra, Synth. Met., 31 (1989) 155.

12- M. Morita, I. Hashida and Masato, J. Appl. Polym. Sc., 36 (1989) 1639.

13- H. Weiss, O. Pfefferkorn, G. Korota and B.D. Humphrey, J. Electrochem. Soc., 136 (1989) 3711.

14- R.V. Gregory, W.C. Kimbrell and H.H. Kuhn, Synth. Met., 28 (1989) C823.

15- E.M. Genies, C. Petrescu and L. Olmedo, ICSM'90.

16- E.M. Genies, French patent, N° 89 08811.

17- a- E.M. Genies, G. Bidan and J.-F. Penneau, J. Electroanal. Chem., 271 (1989) 59. b- A. Andreatta, Y. Cao, J.C. Chiang, A.J. Heeger and P. Smith, Synth. Met., 26 (1988) 383.

18- S. Yang, S.A. Tirmizi, A. Burns, A. Barney and W.M. Risen, jr., Synth. Met., 32 (1989) 191.

19- R. Garreau, J. Roncali, F. Garnier and M. Lemaire, J. Chim. Phys., 86

(1989) 93 and references therein.

20- A. Yassar, J. Roncali and F. Garnier, Polym. Commun. 28 (1987) 103.

  21- S.P. Armes, J.F. Miller, B. Vincent, J. Colloïd and Inter. Sc., 118 (1987) 410.

22- N. Cawdery, T.M. Obey and B. Vincent, J. Chem. Soc., Chem. Commun., (1988) 1189.

23- Y. Ikenoue, N. Uotani, A.O. Patil, F. Wudl and A.J. Heeger, Synth. Met., 30 (1989) 305.

24- E.E. Havinga, L.W. Horsen, W. Hoeve, H. Wynberg and E.W. Meijer, Polym. Bull., 18 (1987) 277.

25- D.D.C. Bradley and Y. Mori, Springer Series in Solid-State Sciences, V.91, "Electronic Properties in Conjugated Polymers III", (1989) 225.

26- P.C. Allen, D.C. Bott, C.S. Brown, L.M. Connors, S. Gray, N.S. Walker, P.I. Clemenson and W.J. Feast, ibid, (1989) 456.

27- D. Delabouglise, J. Roncali, M. Lemaire and F. Garnier, J. Chem. Soc., Chem. Commun., (1989) 475.

28- F. Garnier, this conference.

29- H. Ge and G.G. Wallace, Anal. Chem., 61 (1989) 2391.

30- H. Mao, P.G. Pickup, J. Electroanal Chem., 265 (1989) 127.

31- G. Bidan, B. Ehui and M. Lapkowski, J. Phys.D: Appl. Phys., 21 (1988) 1043.

32- D. Orata and D. Buttry, J. Electroanal. Chem., 257 (1988) 71.

33- S. Holdcroft and B.L. Funt, J. Electroanal. Chem., 240 (1988) 89.

34- C.M. Castro-Acuna, F.F. Fan and A.J. Bard, J. Electroanal. Chem., 234 (1987) 347.

35- P. Enzel and T. Bein, J. Chem. Soc., Chem. Commun., (1989) 1326.

36- A. Epstein, this conference.

37- F. Beck and P. Hülser, J. Electroanal. Chem., 280 (1990) 159.

38- A. Boyle, E.M. Genies and M. Lapkowski, Synth. Met., 28 (1989) 769.

39- C.R. Lowe, N.C. Fouls, S.E. Evans and B.F.Y. Yon Hin, Springer Series in Solid State Sciences, V.91, "Electronic Properties of Conjugated Polymers III", (1989) 432.

40- J.-L. Bredas, this conference.

41- D.T. Bloor, this conference.

42- R. Friend, this conference.

43- S. Roth, this conference.

44- E.M. Genies and F. Miquelino, to be published.

45- B.C. Allen, D.C. Bott, C.S. Brown, L.M. Connors, S. Gray, N.S. Walker,

P.I. Clemensen and W.J. Feast, Springer Series in Solid State Sciences, V.91, "Electronic Properties of Conjugated Polymers III", (1989) 456.

46- E.M. Genies and M. Lapkowski, Synth. Met. 24 (1988) 61.

47- A. Pruss and F. Beck, J. Electroanal. Chem., 172 (1984) 281.

48- P.J. Nigrey, A.G. MacDiarmid and A.J. Heeger, J. Chem. Soc., Chem. Commun., (1979) 594.

49- P. Passiniemi and J.-E. Österholm, Mol. Cryst. Liq. Cryst., 121 (1985) 215.

50- E.M. Genies, P. Hany and C. Santier, J. Appl. Electrochem., 18 (1988) 751.

51- A.J. MacDiarmid, S.-L. Mu, N.L.D. Somasiri and W. Wu, Mol. Cryst. Liq. Cryst., 121 (1985) 187.

52- R. Bittihn, G. Ely, F. Woeffler, Makromol. Chem., Macromol. Symp. 8 (1987) 51.

53- S. Panero, P. Prosperi, B. Klaptse and B. Scrosati, Electrochemica Acta, 31 (1986) 159.

54- T. Nakajima and T. Kawagoe, Synth. Met. 28 (1989) C629.

55- E.M. Genies, A.A. Syed and Tsintavis, Mol. Cryst. Liq. Cryst., 121 (1985) 181.

56- M. Nechtschein, F. Devreux, F. Geboud, E. Vieil, J.M. Pernaut and E. Genies, Synth. Met. 15 (1986) 59.

57- P. Degott, M. Armand and M. Fouletier, Anal. Phys., 93 (1986) 11 and French Patent N° 8314940.

58- A. Périchaud, A. LeMehauté, Eur. Pat. N°860409.

59- E. Genies, Fr. Pat. N°85 18484.

60- M. Armand, private communication.

61- K. Bange and T. Gambke, Adv. Mater., 2 (1990) 10.

62- D. Deroo, 2nd Symposium on Polymer Electrolytes, Sienne (Italy), 14-16 June 1989.

63- C.M. Lampert, Sol. Energy mater., 11 (1984) 1.

64- E.M. Genies, M. Lapkowski, C. Santier, and E. Vieil, Synth. Met. 18 (1987) 631.

65- B. Aurian-Blajeni, S.C. Holleck and B.H. Jackman, J. Appl. Electrochem., 19 (1988) 331.

66- E.M. Genies, S. Langlois and M.-N. Collomb, to be published.

67- F.G.K. Baucke, K. Bange and T. Gambke, Displays 10 1988) 179.

69- F. Cristofini, R. de Surville, M. Jozefowicz, L. Yu and R. Buvet, C.R. Acad. Sci. Paris, 268 (1969) 1346.

Processing of high-performance conducting polymers

A. Andreatta[a,b], S. Tokito[c], J. Moulton[a], P. Smith[a,b,d] and A. J. Heeger[a,b,e]

Institute for Polymers and Organic Solids
University of California at Santa Barbara, Santa Barbara, CA 93106

Abstract

A summary is presented of the electrical and mechanical properties of oriented fibers of poly(3-octylthiophene), POT processed from solution, poly(2,5-di-methoxy-p-phenylene vinylene), PDMPV using precursor-polymer methodology, and from polyaniline, PANI, processed as polyblends with poly(p-phenylene terephthalamide), PPTA, from sulfuric acid. Invariably, a strong correlation was observed between the conductivity and the tensile strength and modulus. We show from basic theoretical concepts that this relationship is an intrinsic feature of conducting polymers.

[a]Materials Department,
[b]UNIAX Corporation, 5375 Overpass Road, Santa Barbara, CA 93111
[c]Permanent address: Department of Materials Science and Technology, Kyushu University, Kasuga-shi, Fukuoka 816, Japan,
[d]Chemical and Nuclear Engineering Department,
[e]Physics Department.

1. INTRODUCTION

Conjugated polymers are of special interest because of the potential of a unique combination of electrical and mechanical properties, particularly if these macromolecules are highly extended and oriented.[1] It has been long recognized, however, that most conjugated polymers tend to be insoluble and infusible because of their chain stiffness. During the past five years, fortunately, significant progress has been made in the development of chemico-physical techniques to process these interesting materials into useful, and oriented articles. For example, the addition of long alkyl side chains to the conjugated main chain[2] has has enabled solution or melt processing. An important subclass of these tractable, conjugated polymers are the poly(3-alkylthiophenes)[2-4]. The latter macro-molecules are soluble in a variety of solvents, meltable, exhibit good conduct-ivities, and are reasonably stable. An additional advantageous feature of this group of conducting polymers is the broad range of both mechanical and conductive properties which may be achieved through variation of the alkyl chain length. Here we demonstrate that through a simple solution processing/drawing route poly(3-octylthiophene) (POT) may be formed into oriented materials, that upon doping display significant improvements in both mechanical and conductive properties with respect to undoped fibers. However, it will be also shown that the relatively bulky side chains decrease the π-electron density (and therewith the carrier density) and the interchain coupling, thus making it more difficult to achieve the structural coherence needed to obtain high carrier mobility and exceptional mechanical properties.

A promising strategy to approach the intrinsic mechanical and electrical properties is through the use of the versatile precursor route which involves the preparation of a processible precursor polymer and subsequent conversion of this material to the conjugated polymer.[5,6] The significant advantage of this route is that the saturated precursor polymers can be processed from solution prior to the thermal conversion to the conjugated final, intractable product. The precursor polymers may, therefore, be drawn prior to and during the thermal conversion process so as to yield oriented, homogeneous conjugated polymers.

Poly(p-phenylenevinylene), PPV, and its derivatives such as poly(di-methyl-p-phenylenevinylene) can be prepared from a precursor polymer, a polyelectrolyte, which is soluble in water.[6] The dimethoxy-derivative of PPV, poly(2,5-dimethoxy-p-phenylene vinylene), PDMPV, has been prepared via a similar precursor route to PPV and exhibited high conductivities after doping. However, the commonly used aqueous solutions of the PDMPV precursor polymer tend to form gels, and the gradual elimination of the sulfonium group in the solid precursor cannot be avoided even at room temperature; both effects make subsequent processing into highly oriented films and fibers difficult. Recently, the Kyushu University group[7] succeeded in the preparation of dense PDMPV film from a new precursor polymer which is soluble in common organic solvents, easily processible, and stable even at 100 °C.

An alternative strategy is to identify stable conjugated polymer systems that can be processed. Of this class, polyaniline (PANI) is certainly a promising example. The use of concentrated acids[8] as solvents for PANI has specific advantages in that both the salt and the base form can be completely dissolved at room temperature, with polymer concentrations of more than 20% (w/w), in concentrated protonic acids such as H_2SO_4, CH_3SO_3H, and CF_3SO_3H. Perhaps more important is the fact when precipitated from acid solution, PANI is obtained in the conducting (protonated) emeraldine salt form.[8] Although the ability to process conducting polyaniline from solution represents genuine progress, fibers and films made from these solutions have mechanical properties which currently are not adequate for many applications, due mainly to the low molecular weight of the polyaniline used.[9] Fibers with significantly enhanced mechanical properties have been obtained by blend processing polyaniline with the rigid chain polymer poly(p-phenylene terephthalamide), PPTA.[10] It is well known that PPTA is processed from solutions in concentrated H_2SO_4 to yield one of the strongest and stiffest fibers commercially available.[11,12]

Finally, it should be noted that even in the area of as-polymerized, intractable conjugated polymers such as polyacetylene progress is continuously made. Improved polymerization and tensile deformation techniques have led to the production of truly high-performance materials with conductivities in the 30-100,000 S/cm range and mechanical properties matching those of high-modulus polymers.

In this review, we present a summary of our recent results on the electrical and mechanical properties of solution-spun POT; fibers made from PDMPV[13] using the precursor polymer methodology; and from PANI using the method of processing as polyblends with PPTA from sulfuric acid.[10]

2. SOLUTION-SPUN POT

Poly(3-octylthiophene) samples were produced with a $FeCl_3$ catalyst and supplied by Dr. J.-E. Österholm of Neste Oy (Finland). The samples designated POT 10 and 113, have weight average molecular weights (M_w) of 89,000 and 189,000, resp. and polydispersities of ~4.

Solutions of a polymer concentration of 10 wt. % in chloroform were wet spun into acetone using a laboratory extrusion device at a speed of 0.19 ml/min through a spinnerette with a diameter of 0.4 mm. The as-spun fibers were wound onto bobbins directly from the coagulation bath and subsequently dried under vacuum for 24 hours. Dark-red fibers with a linear density of approximately 80 denier were obtained.

Drawing of the dried fibers was carried out in a temperature controlled, continuous drawing, tube furnace. It was found that the maximum draw ratio for POT 10 occurred at 105 °C. At this temperature, POT fibers were drawn to ~7 times their original length. The higher molecular weight sample, POT 113, could be drawn to slightly higher draw ratios, although the upper limits were relatively low for both materials. However, even the moderate draw ratios achieved in this work resulted in a reasonable orientation of the POT, as determined by wide angle X-ray scattering.

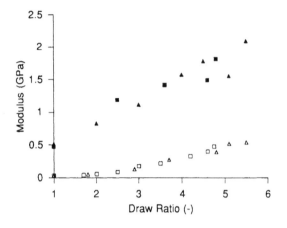

Fig. 1. Initial modulus vs. draw ratio of POT.
□ undoped POT 10,
△ undoped POT 113,
■ FeCl$_3$·6H$_2$O doped POT 10,
▲ FeCl$_3$·6H$_2$O doped POT 113.

Figure 1 shows the Young's modulus of POT fibers versus draw ratio. It is seen that the samples exhibited relatively low moduli, comparable with values for elastomers. Unlike many other conducting polymers, however, the POT samples exhibited substantial increases in both tenacity and modulus when doped to moderate dopant levels, > 18 mole percent, as is also shown in Figure 1. X-ray studies[14] suggest that the doping process involves a spreading, and subsequent interdigitation, of the alkyl side chains. Increased interdigitation may significantly increase the stiffness through the possibility of crystallization between the alkyl side chains.

As with both modulus and tenacity, the electrical conductivity of the doped samples increased linearly with draw ratio.

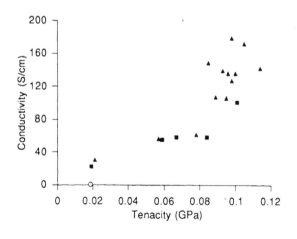

Fig. 2. Conductivity versus tenacity.
▲ FeCl$_3$·6H$_2$O doped POT 113
■ FeCl$_3$·6H$_2$O doped POT 10
○ undoped POT 113.

Figure 2 shows the unique correlation between tensile strength and electrical conductivity of doped POT. A similar relation was observed between the Young's modulus and the conductivity.

3. PRECURSOR ROUTE TO PDMPV

The preparation of the precursor polymer and thermal conversion to the conjugated are presented elsewhere.[13] Thermal analysis of the PDMPV precursor polymer indicated a glass transition at 110 °C, well-separated in temperature from the ~195 °C needed for thermal elimination of the methoxy leaving groups.

The precursor polymer was dissolved into chloroform. The solution was extruded as above at a speed of 0.013 ml/min through a spinneret of a diameter of 0.5 mm into hexane; the resulting precursor fibers were taken up onto a bobbin at a speed of 30 cm/min. The precursor fibers were dried in a vacuum oven overnight; uniform pale yellow fibers were obtained with diameters of 50 - 100 μm. The drawing of the precursor fiber and conversion to fibers of the conjugated polymer were carried out using a temperature controlled, continuous drawing, tube furnace system at a temperature of approximately 200 °C.

The PDMPV fibers were doped by exposure for 1 hr to the vapor pressure (approximately 1 mm Hg) of iodine at room temperature.

The undrawn PDMPV fiber exhibited relatively modest mechanical properties: the modulus and tensile strength were 1.3 GPa and 0.07 GPa, respectively. Figure 3 reveals that fibers which had been drawn to 8 times their initial length had a Young's modulus as high as 35 GPa. The tensile strength was 0.7 GPa. The effect of doping on the modulus is also displayed (filled circles). The data presented in these graphs indicate that doping caused only a moderate reduction of the modulus. Essentially no loss of tensile strength was found.

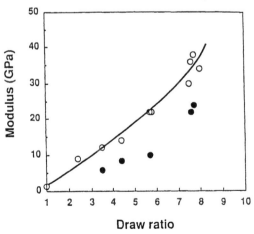

Figure 3: Young's modulus as a function of the draw ratio for undoped (open circles) and for PDMPV fiber doped with iodine (closed circles).

The undrawn PDMPV fiber exhibited a conductivity of 20 S/cm, with a gradual increase of conductivity seen up to a draw ratio of 5. At draw ratios greater than 5, the conductivity increased dramatically, as reported for stretched films[15] (although different in detail, since in ref. 15 the unstretched films were already about 30% converted). At a draw ratio of 8, the conductivity was 1200 S/cm, 60 times higher than that of the undrawn material.

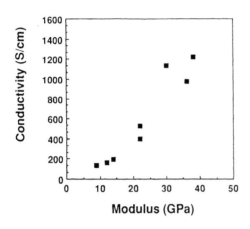

Figure 4: Conductivity as a function of Young's modulus for PDMPV fiber.

Figure 4 shows the relationship between Young's modulus and electrical conductivity for drawn PDMPV fibers. Again, as was previously observed for POT, a strong, essentially linear, correlation was found between conductivity and modulus and (not shown) tenacity.

4. SOLUTION-BLENDED PANI/PPTA

The synthesis of the polyaniline was reported elsewhere[8,9]. Poly(p-phenylene terephthalamide) with inherent viscosity of 7.43 dl/g was obtained from Du Pont as Kevlar® powder.

Homogeneous solutions of PANI and PPTA in 98% H_2SO_4 covering the entire composition range were prepared at room temperature. Since the concentration of PPTA in the solutions was maintained at 1.5 wt% (which is below the onset of the formation of the lyotropic phase; typically at 6-8 wt% PPTA[11,12]), these polyblend fibers were spun from isotropic solutions.

The polymer blend solutions were wet spun into 1N H_2SO_4. Monofilaments were collected onto a take-up spool; tension was applied to elongate the fiber during coagulation. The draw down ratio invariably was as high as possible, for the production of continuous fibers; and it was increased from 7 to 20 with increasing PPTA concentration in the solution. Extrusion speed varied from 0.12 to 0.3 m/min and the wind-up speed from 1.8 to 4.2 m/min. Subsequently, the fibers were washed with running deionized water for 48 hr. The bobbin was then submerged in 1.5N HCl for 12 hr. This allowed the acid to penetrate the fibers and homogeneously protonate them to the conducting emeraldine salt form (partial reduction of the emeraldine salt occurs during the washing; the HCl treatment restores PANI to the fully protonated form). Finally, the bobbin was placed in an oven and the fibers dried under vacuum at 50°C while maintained at constant length by the bobbin. The pure

PPTA fibers were spun with the same method. The pure polyaniline solution was dry-jet wet spun into a 1N H_2SO_4 solution. Washing, HCl treatment, and drying was carried out as described for the blend fibers.

Wide angle X-ray patterns of the composite fibers consisted of superimposed reflections from the two components, indicating that PANI and PPTA segregated during the coagulation.[10]

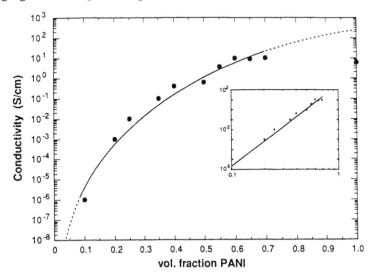

Fig. 5. Conductivity versus volume fraction of PANI in polyblend fibers with PPTA; the inset shows the same data on a log-log plot.

In Figure 5 (inset), the conductivity vs. volume fraction of PANI is shown on a log-log plot. The results indicate that over a relatively wide range of concentrations, the conductivity follows a simple power law,

$$ß = ß_o f^{\alpha} \tag{1}$$

where f is the weight fraction of PANI. The solid curve in Figure 5 shows the power law fit to the experimental data with $\alpha \sim 8$ and $ß_o = 3 \times 10^2$; the conductivity varies with fractional concentration of conducting polymer (from 10-70 % PANI) according to equation 1 over a range of values spanning nearly seven orders of magnitude.

In the context of percolation theory,[16] at sufficiently dilute concentrations such that there are no connected paths, the conductivity would be zero. As the concentration of conducting polymer is increased above the percolation threshold, the conductivity would become finite and increase as the connectivity (i.e. the number of conducting paths) increases. In contrast, the data presented in Figure 5 show no indication of a well-defined percolation threshold. Attempts to estimate a value for the percolation threshold (f_p) by fitting the data to the form $\beta=\beta_0(f-f_p)^\alpha$ lead to the conclusion that f_p is below ~1 %. By contrast, classical percolation theory for a three-dimensional network of conducting polymer globular aggregates in an insulating matrix predicts a percolation threshold at a volume fraction f_p~0.16.[16] Percolation at f_p~0.16 has been observed for conducting polymer composites.[17]

On the other hand, for rod-like structures with length L and diameter d (e.g. a more fibrillar conducting network), the percolation threshold is determined by the excluded volume per fibril (~$\pi L^2 d$).[18] With N fibrils per unit volume, the percolation threshold is given by $f_p \sim N\pi L^2 d = N\pi Ld^2(L/d) = f_{fib}(L/d)$ where f_{fib} is the volume fraction of conducting material in the fibrillar network. Thus, the onset of conductivity would occur at $f_{fib}=f_p(d/L) << f_p$. The analogous percolation of molecularly dispersed rods was observed[19] for polydiacetylene in solution in toluene, where the onset of gelation (i.e. the formation of a connected mechanical network) occurs at a volume fraction of only 0.04%. Since PPTA precipitates from sulfuric acid at a lower water content (at ~10 wt%) than PANI (at ~25 wt%), PPTA comes out of solution first and forms a mechanically connected network. That structure may serve to assemble the conducting PANI along the pre-existing connected paths, thereby leading to the kind of fibrillar conducting pathways that would decrease the percolation threshold to f_p<<0.16.

5. DISCUSSION

Although the electrical conductivity of conducting polymers is enabled by intra-chain transport, in order to avoid the localization inherent to one-dimensional systems, one must have the possibility of interchain charge transfer.[1,20] The electrical transport becomes essentially three-dimensional (and thereby truly metallic) so long as there is a high probability that an electron will have diffused to a neighboring chain between defects on a single chain. For well-ordered crystalline material in which the chains have precise phase order, the interchain diffusion is a coherent process. In this case, the condition for extended transport is that [1,20]

$$L/a >> (t_0/t_{3d}) \qquad (2)$$

where $\underline{L}$ is the coherence length, $\underline{a}$ is the chain repeat unit length, t_0 is the intra-chain π-electron transfer integral, and t_{3d} is the inter-chain π-electron transfer integral. An analogous argument can be constructed for achieving the intrinsic strength of a polymer. If E_0 is the energy required to break the covalent main-chain bond and E_{3d} is the weaker interchain bonding energy (from Van der

Waals forces and hydrogen bonding for saturated polymers), then the requirement is coherence over a length L such that[21,1]

$$L/a \gg E_0/E_{3d}. \qquad (3)$$

In this limit the large number (L/a) of weak interchain bonds add coherently such that the polymer fails by breaking the covalent bond. The direct analogy between equations 2 and 3 is clearly evident. In fact, for conjugated polymers, E_0 results from a combination of ß and π bonds (the latter being equal to t_0, see eqn. 2) and E_{3d} is dominated by the interchain transfer integral, t_{3d}. Thus, equations 2 and 3 predict that quite generally the conductivity and the mechanical properties will improve in a correlated manner as the degree of chain alignment is increased, with each approaching intrinsic values when the inequalities of equations 2 and 3 are satisfied. These predictions are in general agreement with the data obtained for PDMPV as shown in Figure 4. A strong correlation is observed, suggesting that major improvements in electrical conductivity can be anticipated as the materials are further improved such that the mechanical properties approach their intrinsic values. However, until a more quantitative understanding of the implied relationships is attained, extrapolation to the intrinsic electrical conductivity is not possible.

6. CONCLUSIONS

We have shown that a solution processing route for poly(3-octyl-thiophene) results in easily dopable films or fibers, where doping significantly improved in both the mechanical and electrically conductive properties with respect to undoped samples. Conductivities as high as 180 S/cm were obtained for drawn, $FeCl_3$ doped fibers. It was illustrated that oriented PDMPV fibers exhibited excellent mechanical properties (modulus and strength were 35 GPa and 0.7 GPa, respectively) and high electrical conductivity (ß ~1200 S/cm after iodine doping). The mechanical properties were retained after doping; the modulus only decreased to 25 GPa, and the tensile strength remained essentially unchanged. For both polymers strong correlations were observed between mechanical properties and conductivity.

It was also demonstrated that blend fibers of polyaniline and poly(1,4-phenylene terephthalamide) can be derived from solutions in concentrated sulfuric acid. A minor amount of PPTA significantly improved the mechanical properties of PANI fibers while retaining the conductivity of pure polyaniline. These results indicate, therefore, that processing PANI/PPTA from concentrated sulfuric acid represents a viable method for producing polyblend fibers with mechanical properties in the textile range and with moderate levels of electrical conductivity. In this study, the concentration of PPTA was below the onset of the formation of the lyotropic phase; we anticipate that results obtained with lyotropic systems will exhibit significantly improved properties.

We have shown that the modulus and tensile strength derive from a combination of the intra-chain interactions (e.g. strength of chemical bonding, chain conformation, etc.) and inter-chain interaction (e.g. van der Waals forces, interchain transfer interactions, chain conformation, etc.). In conjugated polymers, these same features (band conduction within a polymer chain and efficient electron transfer between polymer chains) determine the carrier mean free path, and thus, the electrical conductivity. Therefore, we conclude that the mechanical and electrical properties of conjugated polymers are intrinsically linked, and we anticipate that in general as the tensile strength (and/or modulus) improve with improved chain orientation, the electrical conductivity will show corresponding improvements until both approach their respective intrinsic theoretical values.

Acknowledgements
The research on PDMPV and POT was supported by the Office of Naval Research (N00014-83-K-0450). The synthesis of the polyaniline used in this study was funded through a MRG grant from the National Science Foundation (NSF–DMR87–03399). The polyblend fiber spinning and the mechanical and electrical measurements were supported jointly by DARPA-AFOSR and monitored by AFOSR under contract no. F49620-88-C-0138.

References:
1. A. J. Heeger, Faraday Discuss. Chem. Soc., 88 (1989) 1.
2. a. K. Y. Jen, R. Oboodi and R. Elsenbaumer, Polym. Mater. Sci. Eng., 53 (1985) 79.
   b. M. J. Nowak, S. D. D. Rughooputh, S. Hotta, and A. J. Heeger, Macromolecules, 20 (1987) 212.
   c. S. D. D. Rughooputh, S. Hotta, A. J. Heeger and F. Wudl, J. Polym. Sci., Polym. Phyd. Ed., 25 (1987) 1071.
   d. M. Sato, S. Tanaka and K. Kaeriyama, J. Chem. Soc. Chem. Commun., 295 (1986) 873.
3. R. L. Elsenbaumer, K. Y. Jen, and R. Oboodi, Synth. Met., 15 (1986) 169.
4. R. Sugimoto, S. Takeda, H. B. Gu, and K. Yoshino, Chem. Express, 1(11) (1986) 635.
5. J. H. Edwards and W. J. Feast, Polym. Commun., 21 (1980) 595.
6. a. D. R. Gagnon, J. D. Capistran, F. E. Karasz and R. W. Lenz, Polym. Bull., 12 (1984) 93.

b. I. Murase, T. Ohnishi and M. Hirooka, Polym. Commun., 25 (1984) 327.

7. T. Momii, S. Tokito, T. Tsutsui and S. Saito, Chem. Lett., (1988) 1201.

8. A. Andreatta, Y. Cao, J. C. Chiang, A. J. Heeger and, P. Smith, Synth. Metals, 26 (1988) 383.

9. Y. Cao, A. Andreatta, A. J. Heeger and P. Smith, Polymer, 30 (1989) 2305.

10. A. Andreatta, A. J. Heeger and P. Smith, Polym. Commun., 31 (1990) 275.

11. H. Blades, U. S. Pat. 3,767,756 (1973), 3,869,429 (1975), 3,869,430 (1975).

12. M. Lewin and J. Preston, ed., *Handbook of Fiber Science and Technology*, Marcel Dekker, Inc., New York , NY, 1985, Vol III, part A, chapter 9.

13. S. Tokito, P. Smith and A. J. Heeger, Polymer, (in press).

14. M. J. Winokur, to be submitted for publication.

15. S. Yamada, S. Tokito, T. Tsutsui and S. Saito, J. Chem. Sci., Chem. Commun., (1987) 1448.

16. R. Zallen, *The Physics of Amorphous Solids*, John Wiley, New York, 1983, Ch.4.

17. S. Hotta, S. D. D. V. Rughooputh and A. J. Heeger, Synth. Met., 22 (1987) 79 .

18. I. Balberg, C. H. Anderson, S. Alexander and N. Wagner, Phys. Rev., B30, (1984) 3933.

19. a. M. Sinclair, K. C. Lim and A. J. Heeger, Phys. Rev. Lett., 53 (1985) 3933.

   b. A. Kapitulnik, K. C. Lim, S. A. Casalnuovo and A. J. Heeger, Macromolecules, 19 (1986) 676.

20. S. Kivelson and A. J. Heeger, Synth. Met., 22 (1988) 371.

21. Y.Termonia and P. Smith, in "The Path to High Modulus Polymers with Stiff and Flexible Chains", Eds A. E. Zachariades and R. S. Porter (Marcel Dekker, New York, 1988) p. 321.

The polyanilines: Potential technology based on new chemistry and new properties

Alan G. MacDiarmid, Department of Chemistry, University of Pennsylvania, Philadelphia, Pennsylvania 19104-6323, U.S.A.
and
Arthur J. Epstein, Department of Physics and Department of Chemistry, The Ohio State University, Columbus, Ohio 43210-1106, U.S.A.

ABSTRACT

The conductivity of doped polyaniline in the emeraldine oxidation state is shown to initially increase with increasing molecular weight and then to remain relatively constant. The processing of the polymer into uniaxially oriented films and fibers and into biaxially oriented films is discussed. The conductivity of uniaxially stretched films and fibers, the apparent degree of crystallinity and the tensile strength of the polymer are shown to increase significantly with the extent to which it is mechanically stretch-oriented. A "self" protonic acid doped form of polyaniline and an interpenetrating polymer network system involving polyaniline, both of which show greatly enhanced conductivity at pH values where doped polyaniline has negligible conductivity are described. Two completely different methods are reported for synthesizing polyaniline in its highest (pernigraniline) oxidation state.

1. INTRODUCTION

The polyanilines are probably the most rapidly growing class of conducting polymers as can be seen from the number of papers and patents published (948) during the last four years, viz, 1986, (108); 1987 (221); 1988 (236); and 1989, (383)[1]. The interest in this conducting polymer stems from the fact that many different ring- and nitrogen- substituted derivatives can be readily synthesized and that each derivative can exist in several different oxidation states which can in principle be "doped" by a variety of different dopants either by non-redox processes or by partial chemical or electrochemical oxidation[2]. These properties, combined with the relative low cost of several polyanilines, their ease of processing and satisfactory environmental stability indicate strongly their significant potential technological applicability.

The polyanilines refer to a class of polymers which can be considered as being derived from a polymer, the base form of which has the generalized composition:

$$\left[\left\{\hspace{-2mm}\bigcirc\hspace{-2mm}\text{-}\overset{H}{N}\text{-}\bigcirc\hspace{-2mm}\text{-}\overset{H}{N}\hspace{-2mm}\right\}_y\left(\bigcirc\hspace{-2mm}\text{-}N=\bigcirc\hspace{-2mm}=N\right)_{1-Y}\right]_x$$ and which consists of alternating

reduced, $-\bigcirc\text{-}\overset{H}{N}\text{-}\bigcirc\text{-}\overset{H}{N}\text{-}$ and oxidized, $-\bigcirc\text{-}N=\bigcirc=N-$ repeat units[3,4]. The

average oxidation state, (1-y) can be varied continuously from zero to give the completely

reduced polymer, $\left\{\bigcirc\text{-}\overset{H}{N}\text{-}\bigcirc\text{-}\overset{H}{N}\text{-}\bigcirc\text{-}\overset{H}{N}\text{-}\bigcirc\text{-}\overset{H}{N}\right\}_x$, to 0.5 to give the "half-

oxidized" polymer, $\left[\left\{\bigcirc\text{-}\overset{H}{N}\text{-}\bigcirc\text{-}\overset{H}{N}\right)\text{-}\left(\bigcirc\text{-}N=\bigcirc=N\right)\right]_x$, to one to give the

completely oxidized polymer, $\left\{\bigcirc\text{-}N=\bigcirc=N\text{-}\bigcirc\text{-}N=\bigcirc=N\right\}_x$.   The terms
"leucoemeraldine", "emeraldine" and "pernigraniline" refer to the different oxidation states
of the polymer where (1-y) = 0, 0.5 and 1 respectively, either in the base form, e.g.
emeraldine base or in the protonated salt form, e.g. emeraldine hydrochloride[2-4]. In
principle, the imine nitrogen atoms can be protonated in whole or in part to give the
corresponding salts, the degree of protonation of the polymeric base depending on its
oxidation state and on the pH of the aqueous acid. Complete protonation of the imine
nitrogen atoms in emeraldine base by, e.g. aqueous HCl, results in the formation of a
delocalized polysemiquinone radical cation[2,4,5] and is accompanied by an increase in
conductivity of $\sim10^{10}$.

2.   SYNTHESIS OF POLYANILINE IN THE EMERALDINE OXIDATION STATE

The partly protonated emeraldine hydrochloride salt can be synthesized easily as a partly
crystalline black-green precipitate (dark green by transmitted light) by the oxidative
polymerization of aniline, $(C_6H_5)NH_2$, in aqueous acid media by a variety of oxidizing
agents, the most commonly used being ammonium peroxydisulfate, $(NH_4)_2S_2O_8$, in
aqueous HCl[2-5]. As noted in Section 4, we have recently shown that the pernigraniline
oxidation state (where (1-y) = 0 in the formula above) is first formed and that this is
subsequently converted to the emeraldine oxidation state. The emeraldine hydrochloride
salt can be deprotonated by aqueous ammonium hydroxide to give an essentially
amorphous black-blue (dark blue by transmitted light) "as-synthesized" emeraldine base
powder with a coppery, metallic glint having an average oxidation state as determined by
volumetric $TiCl_3$ titration corresponding approximately to that of the ideal emeraldine
oxidation state[6]. The ^{13}C [7] and ^{15}N NMR [8] spectra of emeraldine base are consistent with
its being composed principally of alternating oxidized and reduced repeat units.

The emeraldine base form of polyaniline was the first well established example[2,4,9-11] of
the "doping" of an organic polymer to a highly conducting regime by a process in which
the number of electrons associated with the polymer remain unchanged during the doping
process. This was accomplished by treating emeraldine base with aqueous protonic acids
and is accompanied by a 9 to 10 order of magnitude increase in conductivity (to 1 - 5 S/cm;
4 probe; compressed powder pellet) reaching a maximum in ~1M aqueous HCl with the
formation of the fully protonated emeraldine hydrochloride salt, viz.,

$$\left[\left\{\bigcirc\text{-}\overset{H}{N}\text{-}\bigcirc\text{-}\overset{H}{N}\right)\text{-}\left(\bigcirc\text{-}N=\bigcirc=N\right)\right]_x \xrightarrow{2x\ HCl} \left[\left\{\bigcirc\text{-}\overset{H}{N}\text{-}\bigcirc\text{-}\overset{H}{N}\right)\text{-}\left(\bigcirc\text{-}\overset{H}{\underset{\underset{Cl^-}{+}}{N}}\text{-}\bigcirc\text{-}\overset{H}{\underset{\underset{Cl^-}{+}}{N}}\right)\right]_x \quad (1)$$

If the fully protonated i.e. ~50% protonated emeraldine base should have the above dication i.e. bipolaron constitution as shown in equation 1, it would be diamagnetic. However, extensive magnetic studies[2,12] have shown that it is strongly paramagnetic and that its Pauli (temperature independent) magnetic susceptibility increases linearly with the extent of protonation. These observations and other earlier studies[4,9-11] show that the protonated polymer is a polysemiquinone radical cation, one resonance form consisting of two

separated polarons:
It can be seen

from the alternative resonance form where the charge and spin are placed on the other set of nitrogen atoms that the overall structure is expected to have extensive spin and charge delocalization.

3. SELF-PROTONIC ACID DOPED POLYANILINE AND INTERPENETRATING POLYMER NETWORK POLYANILINE.

It is well known that other strong protonic acids such as R-SO$_3$H (R = organic group) besides HCl dope polyaniline according to equation 1. It has recently been found[13] that emeraldine base reacts with fuming H$_2$SO$_4$. A hydrogen atom on the (C$_6$H$_4$) rings is replaced by a -SO$_3$H group thus making R = polyaniline chains. The proton from the -SO$_3$H groups then protonates the -N= groups resulting in a "self-doped" polymer (σ ~0.1S/cm; 4-probe; compressed pellet) in which the protonic acid (dopant) is part of the polymer itself!, viz.,

This polymer differs remarkably from

conventionally protonated emeraldine base in that its conductivity remains unchanged when equilibrated with aqueous acid in the pH range 0 to 7. Emeraldine hydrochloride for example is essentially deprotonated even at pH 4-5 where its conductivity is very small; at pH 7 it is an insulator[4]. The concentration of the -SO$_3^-$ groups and hence of the protons in the "volume" of a chain is ~1-10 molar. This is 8-9 orders of magnitude higher than the proton concentration of a neutral aqueous solution of pH = 7 with which, for example, it could be in equilibrium. Electrostatic interactions prevent the proton from diffusing into the solution away from the negatively-charged polymer chain. This "enhanced proton concentration effect" is responsible for the high conductivity even when the polymer is in equilibrium with a mild acid or neutral solution. When treated with, for example, aqueous NaOH solution, the polymer readily dissolves forming the corresponding sodium salt,

which is an insulator.

A very similar effect has been observed in a self-protonic acid doped interpenetrating polymer network system formed from polyaniline (emeraldine oxidation state) and the polymeric dopant acid, polyvinyl sulfonic acid, [-CH$_2$-CH(SO$_3$H)-]$_x$[14]. The synthesis of the material is carried out in exactly the same manner as the synthesis of emeraldine hydrochloride from aniline except that polyvinyl sulfonic acid is substituted for HCl. The polyvinyl sulfonic acid is actually formed *in situ* by using equimolar quantities of sulfonate groups (from the sodium salt of the polyvinyl sulfonic acid) and HCl. Elemental analyses of the product show that only negligible amounts of chlorine are incorporated into the product which is formed as precipitate and also as a strongly adherent film on surfaces of certain substrates such as indium oxide coated glass placed in the reaction mixture during the polymerization process. The conductivity of the dried powder after equilibration in buffer solutions falls slowly from 1 to 0.1 S/cm as the pH increases from 0 to 7 and then falls rapidly to a value $<10^{-8}$ S/cm at pH=8. The interpenetrating polymer network of polyaniline and polyvinyl sulfonic acid chains formed during the polymerization process does not become disentangled up to a pH of 7 during the above treatment. This results in a local effective proton concentration many orders of magnitude higher than that of the surrounding aqueous medium. Electronic and infrared studies are consistent with the above conductivity/pH results.

4.   SYNTHESIS OF THE PERNIGRANILINE OXIDATION STATE

Pure pernigraniline, the completely oxidized form of polyaniline,

$$\left[\langle\!\!\!\!\!\bigcirc\!\!\!\!\!\rangle\text{—N}=\!\!\!\langle\!\!\!\!\!\bigcirc\!\!\!\!\!\rangle\!=\!\text{N}\text{—}\langle\!\!\!\!\!\bigcirc\!\!\!\!\!\rangle\text{—N}=\!\!\!\langle\!\!\!\!\!\bigcirc\!\!\!\!\!\rangle\!=\!\text{N} \right]_x ,$$ has recently been synthesized for the first time.

It was reported in 1910[15] to be formed in an impure state by the oxidation of emeraldine base but that it was unstable and rapidly decomposed, especially when wet. Synthesis of the analytically pure, dark purple, partially crystalline powder has been accomplished by the controlled oxidation of emeraldine base with $m\text{-Cl}(C_6H_4)C(O)OOH/N(C_2H_5)_3$ in NMP[16]. Free-standing, lustrous, copper-colored films can be cast from this solution and subsequently leached in a methanol/acetone mixture to remove excess oxidizing agent.

We have found recently[17], using a potential profiling technique whereby the potential of the system in which the aniline is undergoing polymerization is constantly monitored, that polyaniline in the pernigraniline oxidation state is the first formed product in the common method of synthesizing the emeraldine oxidation state by the oxidative polymerization of aniline in aq. 1.0M HCl by $(NH_4)_2S_2O_8$ [17]. This is the case both for the *in-situ* deposition of polyaniline as thin films on substrates placed in the polymerization solution and for the synthesis of the polymer as a bulk powder.

When the synthesis is carried out at ~0 °C using excess aniline[17], the initial oxidation potential of the reaction system increases from ~0.40V (vs. SCE) (Figure 1, point A) to ~0.66V (point B) immediately after adding the $(NH_4)_2S_2O_8$ to ~0.75V (point C) within ~2 minutes at which value it stays essentially constant before it begins to fall rapidly after ~10 minutes (point D), reaching ~0.47V at point E, and ~0.44V at point F, a value characteristic of the emeraldine oxidation state[18,19]. The temperature increases noticeably, on going from point D to point E. The above times and potentials are, of course, dependant on the temperature at which the reaction is studied. If the reaction mixture at point D is poured into cold (~5 °C) aqueous NaOH solution, analytically pure pernigraniline base powder which has an oxidation state (1-y) = 0.96±0.02 by $TiCl_3$ titration is obtained. Its electronic and infrared spectra are identical to those of pernigraniline base powder synthesized using $m\text{-Cl}(C_6H_4)C(O)OOH/N(C_2H_5)_3$ as described above.

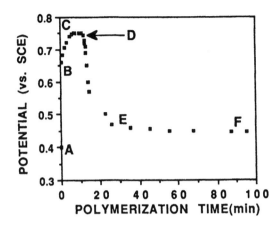

Figure 1. Potential-Time Profile of a Conventional Chemical Oxidative Polymerization of Aniline Using Ammonium Peroxydisulfate in aq. 1.0M HCl (Pt electrode; SCE reference).

At point D, the $(NH_4)_2S_2O_8$ (whose concentration was monitored continuously with time and determined quantitatively by a titrimetric procedure) was essentially all consumed. If the reaction mixture was poured into aqueous NH_4OH at point F, a precipitate of analytically pure emeraldine base was obtained. When the pure pernigraniline base, formed at point D was treated with an excess of a mixture of aniline and HCl of the same relative concentration as that employed in the initial polymerization reaction (but with the omission of $(NH_4)_2S_2O_8$), analytically pure emeraldine base was obtained.

The above observations are consistent with the following hypothesis: The potential at which aniline undergoes electrochemical oxidation in aqueous acid is ~0.7V (vs. SCE)[18]. The potential of the 0.1M solution of $(NH_4)_2S_2O_8$ in 1.0M aqueous HCl before adding aniline was ~1.05V and that of pernigraniline and emeraldine bases in 1M HCl are ~0.83V[16] and ~0.43V[19] respectively. The $(NH_4)_2S_2O_8/$ HCl system is therefore expected to produce, at least initially, polyaniline in its highest (pernigraniline) oxidation state. The pernigraniline oxidation state is a sufficiently strong oxidizing agent to oxidatively polymerize any excess aniline to the emeraldine oxidation state, while it is itself also reduced to the emeraldine oxidation state. The emeraldine oxidation state finally obtained as point F is approached is therefore formed in two different ways: (i) by the reduction of the pernigraniline initially formed and (ii) by the oxidative polymerization of aniline by pernigraniline.

5. MOLECULAR WEIGHT OF POLYANILINE IN THE EMERALDINE OXIDATION STATE

As-synthesized[5] emeraldine base in NMP solution, after passage through a 5000Å micropore filter exhibits a bimodal molecular weight distribution curve by G.P.C. (polystyrene standard) which consists primarily (~85%) of polymer having a peak molecular weight of ~34,000 together with a smaller amount (~15%) of material having a peak molecular weight of ~970,000 [2,20,21]. However, when dissolved in NMP containing 0.5% weight percent LiCl, which apparently acts as a "better" solvent, the higher molecular weight fraction is absent. The monomodal symmetrical peak obtained, gives values of $M_p = 38,000$, $M_w = 78,000$ and $M_n = 26,000$; polydispersity, $M_w/M_n = $ ~3.0. It is concluded that the higher molecular weight fraction actually consisted

Table I
Characterization of Fractions of Emeraldine Base
From Preparative G.P.C. Studies

	M_p	M_n	M_w	M_w/M_n	Conductivity (a) (S/cm)
Fraction 1	15,000	12,000	22,000	1.8	1.2
2	29,000	22,000	42,000	1.9	2.4
3	58,000	40,000	73,000	1.8	7.9
4	96,000	78,000	125,000	1.6	13.1
5	174,000	148,000	211,000	1.4	17.0
6	320,000	264,000	380,000	1.4	14.9

(a) Compressed pellet; 4-probe; after doping with 1M aq. HCl for 48 hours.

of small particles of the polymer which passed through the 5,000Å micropore filter and which were dissolved by the NMP/LiCl solvent. A solution of emeraldine base in NMP/LiCl was then passed through a preparative G.P.C. column and six separate fractions were collected, the lowest molecular weight fraction ($M_p < 5000$) being discarded, since in a separate study it was shown that it contained oxygen-containing impurities[20]. This procedure was repeated 20 times in order to obtain sufficient material for further studies. Each of the six fractions (Table 1) were shown to be pure emeraldine base by elemental analysis, infrared and electronic spectral studies and by cyclic voltammetry[22].

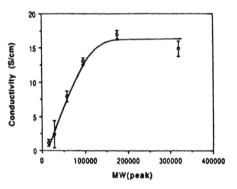

Figure 2. Dependency of Conductivity of Doped Polyaniline (Emeraldine Oxidation State) on Molecular Weight.

It can be seen from Figure 2 that the conductivity of the doped (1M HCl) polymer rises monotonically with molecular weight up to a value of ~150,000 (~1,600 ring-nitrogen repeat units) after which it changes relatively little. It should be noted that the above molecular weights are relative to polystyrene and are not absolute values. The reason for the change in dependency of conductivity on molecular weight is not clearly apparent; however, it is not caused by a change in the degree of crystallinity, since all fractions exhibited approximately the same crystallinity by x-ray diffraction studies.

6.   ORIENTED FILMS AND FIBERS OF POLYANILINE

"As-synthesized" emeraldine base is soluble in N-methyl pyrrolidinone, NMP[23]. However, the term "soluble" must be used with caution since it is not clear how much of the polymer in, for example, a viscous ~20% by weight "solution" is in "true" solution[23,24,25]. It has been known for some time that emeraldine base is readily solution-processible[23,25] and that it may be cast as free-standing, flexible, coppery-colored films from its solutions in NMP. These films can be doped with ~1M aqueous HCl to give the corresponding flexible, lustrous, purple-blue films (σ ~1-4 S/cm) of emeraldine hydrochloride[23] which are partly crystalline[2].

Uniaxially oriented, partly crystalline emeraldine base films are obtained by simultaneous heat treatment and mechanical stretching of films formed from "as-synthesized" emeraldine base containing ~15% by weight plasticizer such as NMP[2,26]. Samples are observed to elongate by up to four times their original length when stretched above the glass transition temperature [> ~110 °C][2,26]. The resulting films have an anisotropic X-ray diffraction and optical response, with a misorientation of only a few degrees[2].

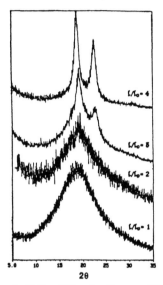

Lustrous, copper-colored ribbons of uniaxially oriented emeraldine base film up to 1.2 meters (4 feet) in length and 2.5 cm in width (thickness ~20μm) of various draw ratios can be readily fabricated by stretch-orienting emeraldine base films cast from NMP solution at ~140 °C between two metal rollers rotating at different speeds[27]. As can be seen from Figure 3 the apparent degree of crystallinity is greatly increased by processing of this type.

Figure 3. X-Ray Diffraction Spectra of Ribbons of Emeraldine Base of Increasing Draw Ratio, (l/l_0; l = final length; l_0 = original length before stretching.)

The tensile strength of the ribbons also increases significantly with an increase in draw ratio (and crystallinity) as shown in Table II. As expected, biaxially oriented film exhibits significantly greater tensile strength than uniaxially oriented film for the same draw ratio (l/l_0 = 2) (Table II). The conductivity of the HCl-doped uniaxially oriented ribbons increases on stretching (l/l_0 = 1, σ ~5 S/cm; l/l_0 = 4, σ ~80 S/cm). It should be noted that the conductivity of the oriented films is greatly dependent on their method of drying; conductivities of ~400 S/cm can be obtained for films which have not been dried to any great extent[27].

The above observations show that polyaniline can be processed by methods used for commercial polymers. Even at this very early stage, its tensile strength overlaps the lower tensile strength range of commercial polymers such as Nylon 6 (Unstretched Nylon 6 has tensile strengths ranging from 69 to 81 MPa[28]).

Table II
Tensile Strength (MPa)[a] of Emeraldine Base Ribbons
as a Function of Draw Ratio (l/l_o)

	Uniaxial Orientation:				Biaxial Orientation[b]:
	l/l_o=1	l/l_o=2	l/l_o=3	l/l_o=4	l/l_o=2
Tensile Strength(Av.)	54.4	53.2	75.9	124.1	122.4
(Best)	59.9	62.1	82.8	144.8	131.6

(a) gauge length = 3 inches   (b) l/l_o = 2 in both directions

Fibers (~30-70μm) of emeraldine base can be formed by drawing a ~20% by weight "solution" of emeraldine base in NMP in a water/NMP solution[24,29]. If desired, the emeraldine base "solution" in NMP may also be drawn in aqueous HCl which results in direct formation of the doped fiber. Fibers can also be spun from NMP solution. The drawn fibers (containing NMP as plasticizer) can be thermally stretched oriented at ~140 °C up to four times their original length in a similar manner to emeraldine base films[24]. X-ray diffraction studies show directional enhancement of the Debye-Scherrer rings. The change in X-ray diffraction spectra of stretch-oriented fibers after doping in 1M HCl are shown in Figure 4. A monotonic increase in apparent crystallinity with draw ratio is observed. Doping with 1M aqueous HCl results in a significant increase in the conductivity parallel to the direction of stretching (σ ~40-170 S/cm) as compared to the conductivity of the polymer powder from which the fibers are prepared (σ ~1-5 S/cm).

As can be seen from Figure 5, the conductivity of the HCl-doped drawn fibers increases monotonically with draw ratio, (l/l_o). It should be stressed that the above data were obtained using "as-synthesized" emeraldine base containing low molecular weight polymer and low molecular weight impurities[20]. Since l/l_o in general will increase with increasing molecular weight, it is apparent that use of higher molecular polymer should result in greater l/l_o ratios and hence in even higher conductivities. The above draw ratio, crystallinity and conductivity data show that the conductivity of the fibers increases monotonically with increase in apparent crystallinity.

Preliminary studies show that emeraldine base fibers both before and after doping

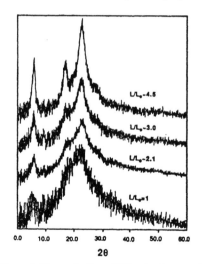

Figure 4. Change in X-ray Diffraction Spectra of Stretch-Oriented Fibers of Emeraldine Base Drawn in Aqueous/NMP to Selected Values After Doping in 1M HCl, (l_o = original length, l = final length).

with 1M aqueous HCl exhibit promising mechanical properties[24]. Values for one inch gauge length (tensile strength, MPa; initial modulus, GPa) for emeraldine base fibers stretch-oriented ($l/l_0 \sim$ 3-4) at ~140 °C after drawing are: 318(Av.); 366(Best) and 8.1(Av.); 8.6(Best). After doping, corresponding values are: 150(Av.); 176(Best) and 4.6(Av.); 5.0(Best). X-ray diffraction studies show some reduction in crystallinity after doping, consistent with the reduction in tensile strength. As expected, the tensile strength of the oriented emeraldine base fibers are greater than those of oriented films. As can be seen on comparing the above tensile strengths with those of, for example, fibers of Nylon 6 (200-905 MPa)[30] the mechanical properties of polyaniline fibers, considering the early stage of development, are most encouraging.

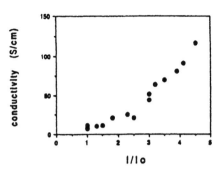

Figure 5. Conductivity of the HCl-doped Drawn Fibers vs. Draw Ratio (l/l_0).

7. CONCLUSIONS

The greater understanding and control of the synthesis of pure polyaniline in selected oxidation states obtained in the present study together with the improved ability to process the pure polymer by conventional methods in the form of oriented films and fibers having greatly increased conductivities and having tensile strengths overlapping those of conventional commercial polymers is most encouraging from the point of view of its technological potential. The data obtained show that its intrinsic properties can only be approached through processing and that increase in its conductivity and mechanical properties are controlled by the same fundamental factors, principally, molecular weight and degree of crystallinity.

8. ACKNOWLEDGEMENTS

The authors are particularly indebted to Mr. Xun Tang and to Dr. Rakesh Kohli for their untiring efforts in assisting with the preparation of this manuscript. The recent work described in this report was supported primarily by the Defense Advanced Research Projects Agency through a contract administered by the Office of Naval Research and in part by N.S.F. Grant No. DMR-88-19885.

9. REFERENCES

1. Based on search of STN International On Line Computer File, Chem. Abs., Aug. 9, 1990.

2. A.G. MacDiarmid and A.J. Epstein 1989 *Faraday Discuss. Chem. Soc.* **88** 317.

3. J-C. Chiang and A.G. MacDiarmid 1986 *Synth. Met.* **13** 193 .

4. A.G. MacDiarmid, J-C. Chiang, A.F. Richter and A.J. Epstein 1987 *Synth. Met.* **18** 285.

5. A.G. MacDiarmid, J-C. Chiang, A.F. Richter, N.L.D. Somasiri and A.J. Epstein 1987 *Conducting Polymers* ed. L. Alcacér (Reidel Publications, Dordrecht), pp105.

6. G.E. Asturias, R.P. McCall, A.G. MacDiarmid and A.J. Epstein 1989 *Synth. Met.* **29** E157.

7. S. Kaplan, E.M. Conwell, A.F. Richter and A.G. MacDiarmid 1989 *Synth. Met.* **29** E235.

8. A.F. Richter, A. Ray, K.V. Ramanathan, S.K. Manohar, G.T. Furst, S.J. Opella, A.G. MacDiarmid and A.J. Epstein 1989 *Synth. Met.* **29** E243.

9. A.J. Epstein, J.M. Ginder, F. Zuo, H-S. Woo, D.B. Tanner, A.F. Richter, M. Angelopoulos, W-S. Huang and A.G. MacDiarmid 1987 *Synth. Met.* **21** 63.

10. P.M. McManus, S.C. Yang and R.J. Cushman 1985 *J. Chem. Soc., Chem. Commun.* 1556.

11. G.E. Wnek 1986 *Synth. Met.* **15** 213.

12. A.J. Epstein, J.M. Ginder, F. Zuo, R.W. Bigelow, H.S. Woo, D.B. Tanner, A.F. Richter, W-S. Huang and A.G. MacDiarmid 1987 *Synth. Met.* **18** 303; J.M. Ginder, A.F. Richter, A.G. MacDiarmid and A.J. Epstein 1987 *Solid State Commun.* **63** 97.

13. J. Yue and A. J. Epstein 1990 *J.Am.Chem.Soc.* **112** 2800.

14. G.E. Asturias, A.G. MacDiamid and A.J. Epstein 1990 unpublished observation.

15. A.G. Green and A.E. Woodhead 1910 *J. Chem. Soc. Trans.* **97** 2388; A.G. Green and A.E. Woodhead 1912 *J. Chem. Soc. Trans.* **101** 1117.

16. Y. Sun, A. G. MacDiarmid and A. J. Epstein 1990 *J. Chem. Soc., Chem. Commun.* 529.

17. S.K. Manohar, A.G. MacDiarmid and A.J. Epstein 1989 *Bull. Am. Phys. Soc.* **34** 582; A.G. MacDiarmid and A.J. Epstein 1990 *Conjugated Polymeric Materials: Opportunities in Electronics, Optoelectronics, and Molecular Electronics* eds. J.L. Brédas and R.R. Chance (Kluwer Academic Publishers, Netherlands) pp53.

18. W-S. Huang, B.D. Humphrey and A.G. MacDiarmid 1986 *J. Chem. Soc., Faraday Trans.* **82** 2385.

19. W-S. Huang, A.G. MacDiarmid and A.J. Epstein 1987 *J. Chem. Soc., Chem. Commun.*, 1784.

20. A.G. MacDiarmid, G.E. Asturias, D.L. Kershner, S.K. Manohar, A. Ray, E.M. Scherr, Y. Sun, X. Tang and A.J. Epstein 1989 *Polym. Prepr.* **30**-(1) 147.

21. X. Tang, Y. Sun and Y. Wei 1988 *Makromol. Chem., Rapid Commun.* **9** 829.

22. X. Tang, A.G. MacDiarmid and A.J. Epstein 1990 unpublished observations.

23. M. Angelopoulos, G.E. Asturias, S.P. Ermer, A. Ray, G.M. Scherr, A.G. MacDiarmid, M. Akhtar, Z. Kiss and A.J. Epstein 1988 *Mol. Cryst. Liq. Cryst.* **160** 151.

24. X. Tang, A.G. MacDiarmid and A.J. Epstein 1989 unpublished Observations.

25. M. Angelopoulos, A. Ray, A.G. MacDiarmid and A.J. Epstein 1987 *Synth. Met.* **21** 21.

26. K.R. Cromack, M.E. Jozefowicz, J.M. Ginder, R.P. McCall, A.J. Epstein, E.M. Scherr and A.G. MacDiarmid 1989 *Bull. Am. Phys. Soc.* **34**: 583.

27. E.M. Scherr, A.G. MacDiarmid, Z. Wang, A.J. Epstein, M.A. Druy, and P.J. Glatkowski 1990 unpublished observations.

28. J. Brandrup and E.H. Immergut 1975 *Polymer Handbook* (John Wiley and Sons, New York), pp 2 (Ch. VIII).

29. X. Tang, E.M. Scherr, A.G. MacDiarmid and A.J. Epstein 1989 *Bull. Am. Phys. Soc.* **34** 583.

30. "Textile World Man-Made Fiber Chart," McGraw-Hill 1988.

Conductive polymers in molecular electronics: Conductivity and photoconductivity

SIEGMAR ROTH

1. INTRODUCTION

Conductive polymers has been a rapidly expanding field of cross-disciplinary research (Roth and Filzmoser 1990) ever since the discovery that polyacetylene will become electrically conducting after treatments with oxidizing or reducing agents (Chiang et al. 1977). A topical conference just now being organized for September 1990 in Tübingen will have to deal with as much as 800 original scientific contributions (Hanack and Roth 1990). Recent review papers have been published amongst others by Roth and Filzmoser (1990), Yu Lu (1988), Heeger et al. (1988), Roth and Bleier (1987), and Skotheim (1986). One of the latest achievements is the synthesis of polyacetylene samples with conductivities after doping in excess of 10^5 S/cm, i.e. nearly as high as that of copper at room temperature (Naarmann and Theophilou 1987, Tsukamoto et al. 1990).

The essential chemical feature of polyacetylene and of the other conductive polymers is the existence of extended systems of conjugated double bonds and it is generally accepted that along the polymer chain these substances are "one-dimensional metals", which - at least in theory - exhibit a variety of typical one-dimensional phenomena such as a Peierls transition, charge density waves (or bond order waves), and non-linear solitonic or polaronic excitations, all of them thoroughly discussed in the review articles and monographs mentioned above and in the original papers cited therein. There is little surprise that also the question has been asked whether a polymer chain with conjugated double bonds could be regarded as a "molecular wire" and whether such wires could be used to connect "molecular electronic devices" or whether a segment of such a chain would act as an electronic device itself. So Carter (1982) has coined the term "soliton switching" and subsequently many speculations have been performed along these lines. Some overview is given in the proceedings of the International Conference on "Molecular Electronics - Science and Technology" (Aviram 1989).

In Fig. 1 the chemical structure of a polyacetylene chain is shown. At selected points the side groups R_1 to R_4 are attached. There are several ways of interpreting the figure:

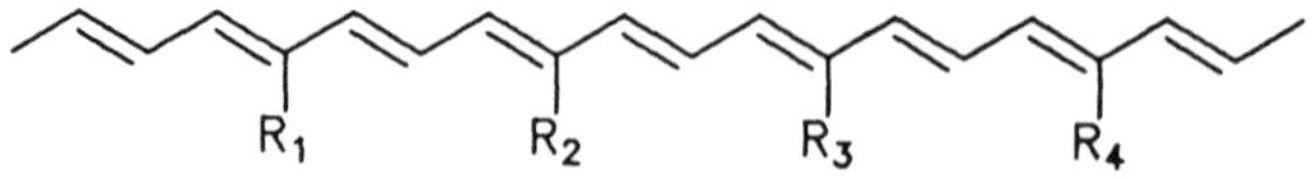

Fig. 1.   Chemical structure of polyacetylene (the prototype of conducting polymers) with functional side groups R_1 to R_4

i)   The side groups could be point contacts and thus the diagram could indicate a set-up for measuring the electrical conductivity of a single polymer chain.

ii) If the side groups were to carry a magnetic moment these moments could interact via the conjugated bonds and the molecule would be a model of an organic ferromagnet (Ovchinnikov et al. 1988, Friend 1987).

iii) By implementing logical input and output units as sidegroups (which might be addressed optically) the molecule would function as an element of molecular electronics.

We know, of course, what to expect from electrical measurements on a monomolecular chains. Because of the one-dimensionality of the system the wavefunctions will be localized (Thoughless 1977) and the resistance will be no longer proportional to the length of the sample so that the concept of resistivity (or conductivity) breaks down. The study of "quantum wires" has shown (Landauer 1970) that in the one-dimensional limit the conductance will be independent of the sample length and quantized in integers of e^2/h (Joachim 1990). The measured electrical conductivity of conducting polymers comes about from the non-negligible coupling between the polymer chains which suppresses the localization. A study of the necessary amount of the three-dimensional coupling is, for example, given in Heeger's (1989) article on high-performance conjugated polymers.

2. CONDUCTIVITY

In the chart of Fig. 2 the conductivity that can be obtained in conjugated polymers is compared to values of some more conventional solids. Undoped conjugated polymers are insulators. They become conducting only upon doping. The physics of the insulator-to-metal transition is discussed in most of the review articles cited in the introduction. It should be noted that the conductivity range between diamond and copper (nearly!) can be spanned even twice, both for p-doping (oxidative doping) and for n-doping (reductive-doping). The hatched part of the arrow in Fig. 2 corresponds to the "new polyacetylenes" (Naarmann and Theophilou 1987 and Tsukamoto et al. 1990). The highest values published so far are about $5 \cdot 10^6$ S/cm (that of copper is $6 \cdot 10^6$ S/cm), but an accurate determination is difficult because of the uncertainties in the sample dimensions (usually very thin films, not more than a couple of microns thick).

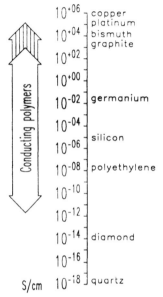

Fig. 2. Conductivity chart

So far it has not been possible to synthesize polyacetylene with a conductivity in excess of 10^5 S/cm by other groups than those of the above mentioned authors. In our laboratory in Stuttgart we have repeated both synthetical routes and with some care we succeed in getting conductivities of about 20000 S/cm along the stretch direction. The temperature dependence of the conductivity of such a sample, parallel and perpendicular to the stretch directions, is shown in Fig. 3. As can be seen, there is always a positive temperature coefficient (opposite to the behaviour of conventional metals). This positive coefficient is even observed in Naarmann's and Tsukamoto's samples. This behaviour is interpreted by the assumption, that the conductivity is still limited by sample imperfection and that the "intrinsic" conductivity is considerably higher. Theoretical estimates of the intrinsic conductivity of polyacetylene are in the order of 10^7 S/cm (Pietronero 1983, Kivelson and Heeger 1988).

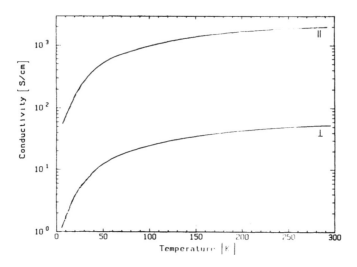

Fig. 3. Temperature dependence of the electrical conductivity of stretch-oriented highly conducting polyacetylene, parallel and perpendicular to the stretch direction

3. PHOTOCONDUCTIVITY

In the context of molecular electronics the photoconductivity is probably more relevant than the dark conductivity. The photoconductivity is a property of the undoped conjugated polymer. In Fig. 4 the experimental set-up for measuring the transient photoconductivity is sketched. The polymer sample is placed into a fairly high

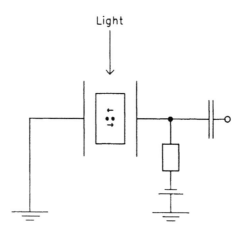

Fig. 4. Experimental set-up for measuring the transient photoconductivity in conjugated polymers

electric field and then illuminated by short light pulses (of ps length). The electric field separates the photogenerated charge carriers and a photocurrent flows till the carriers hit the electrodes or vanish in traps. The photoconductive signal is registered as voltage drop across the load resistor. A typical signal trace is shown in Fig. 5 (Yacoby et al. 1985, Bleier et al. 1987 and 1988). The observed signal decays within some 100 ps. Since this is about the time resolution of the electronic system (fast oscilloscope) the actual photoconductivity could decay even faster. Thus the sample in Fig. 4 acts as a fast optical switch and would be an electronic device based on a molecular material (although not a device treating information on a molecular level). Yet in Fig. 4 not the device aspect is important but the measurement of the pyhsical properties of the conjugated polymer. From the integral over the signal the distance can be calculated which the charge carriers move (Schubweg). With reasonable assumptions for the quantum efficiency in a field of $1.5 \cdot 10^4$ V/cm a Schubweg of 400 Å is found, indicating that it is possible in these systems to move charge carriers quickly over considerable distances. (In addition to the fast component of the photoconductivity shown in Fig. 5 there is also a slow component not visible at the scale of the figure and not relevant for molecular electronics.)

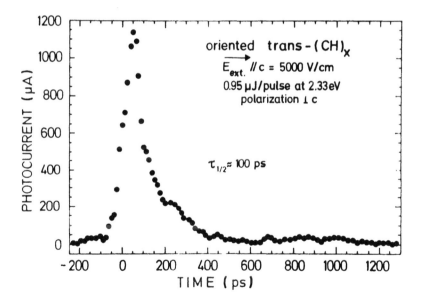

Fig. 5. Transient photoconductivity of trans-polyacetylene

4. MOLECULAR ELECTRONICS

In present-day microelectronics there is a trend towards a continuous decrease of the lateral dimensions of the functional elements. If this trend persists for some decades the structure sizes will be in the order of atomic distances. Then it might turn out that it is easier to assemble electronic structures atom by atom rather than carving them out of a large monolithic block. This is the concept of molecular electronics: "small up" instead of "large down". It is not at all evident that this strategy will be technologically advantageous. One route of small up would be synthetic organic chemistry and one proposal for molecular electronics is the use of conjugated double bonds (Carter 1982).

Nature has furnished an existence proof for information treatment on a molecular level using polyenes (polyenes is the chemical name for chain-like molecules with conjugated double bonds): light absorption of retinal as primary step of optical vision in our eyes. In Fig. 6 a molecule of 11-cis-Retinal is shown. Due to the cis conformation the molecule is bent. After light absorption is stretches into the all-trans form. This stretching triggers a sequence of processes which finally leads to a signal travelling from our eye to our brain. An existence proof for a memory on a molecular level is the genetic code of DNA.

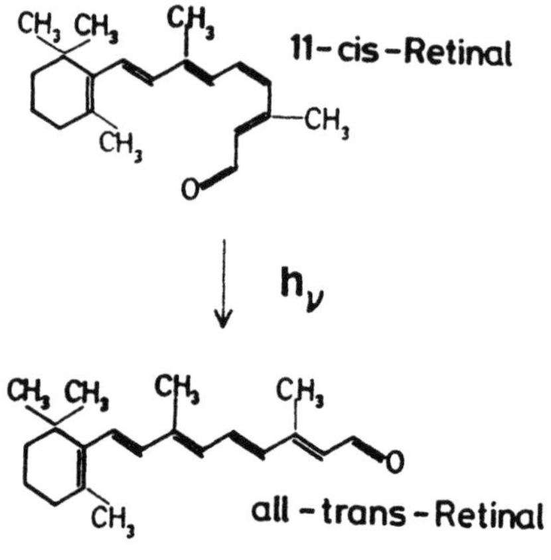

Fig. 6. A molecule of 11-cis-Retinal, the light-sensitive device in optical vision

Carter's proposal of soliton switching is based on mobile domain walls in polyenes. Fig. 7 shows a piece of a polyacetylene chain where two differently conjugated domains are seperated by a domain wall. This domain wall is the famous soliton, which

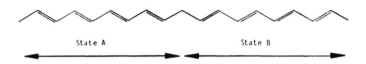

State A   State B

Fig. 7. Polyacetylene chain with two differently conjugated segments separated by a domain wall ("soliton")

has been extensively discussed in the polyacetylene literature (see references above). If the soliton moves along the chain it switches from State A to State B. In very long molecules State A and State B are degenerate (they have the same energy). In small molecules the question arises, how well defined are these states? This will strongly depend on the end groups and in assymetric molecules one state is favoured. If the other (metastable) state is living sufficiently long the molecule could be switched between the two states by the absorption of light. In Fig. 8 a polyene chain with donor and acceptor end groups is shown. Such molecules are known from dye chemistry. In the context of molecular electronics they have been synthesized by several groups (Göhring et al. 1987, Effenberger et al. 1988, Li et al. 1989, Barzoukas et al. 1989). In Fig. 8 the two states do not only differ by the bond arrangement but also by the dipole moment. (Light absorption activates the donor and the acceptor and in the excited state the dipole moment is considerably larger than in the ground state. Because of this light-induced dipole change the molecules are also interesting for non-linear optics.)

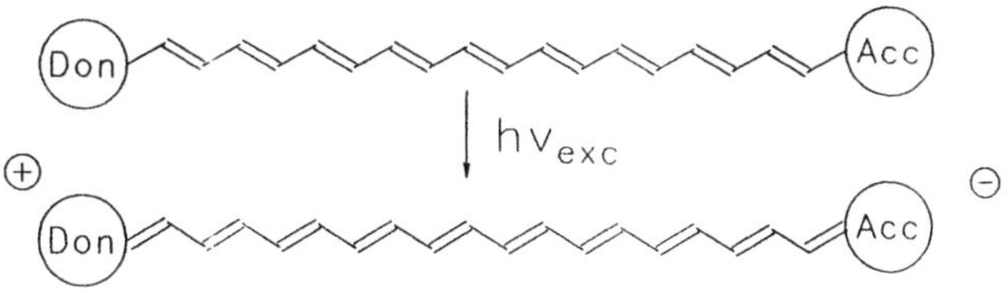

(Don) ⌇⌇⌇⌇⌇⌇⌇⌇⌇⌇ (Acc)

$h\nu_{exc}$

(+)

(Don) ⌇⌇⌇⌇⌇⌇⌇⌇⌇⌇ (Acc)   (−)

Fig. 8. Polyene chain with donor and acceptor end groups

Formally the transition from the ground state to the excited state can be described by the exchanges of solitons between the end groups (Roth 1986). Usually the donor and the acceptor are so strongly coupled that one cannot expect to excite the molecule on one side and watch the arrival of the excitation on the other end. In order to decouple the functional groups at the chain ends one can insert a "spacer" into the chain, e.g. a non-conjugated chain segment as indicated in Fig. 9. Now it will take some time (on the average) till the excitation can tunnel through the spacer. Such molecules with spacer have already been synthesized in our laboratory in Stuttgart and presently we are in the process of characterizing them.

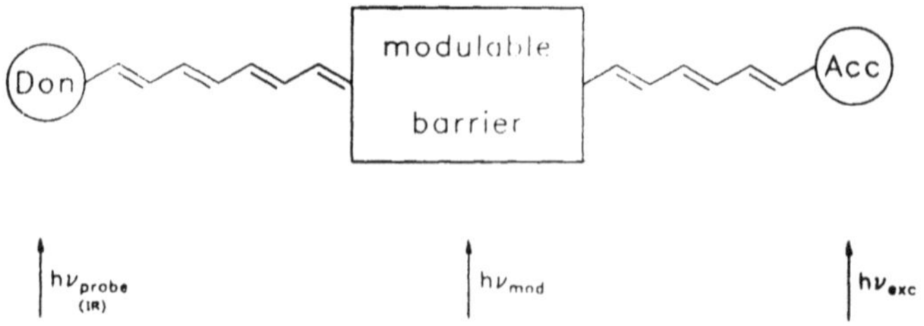

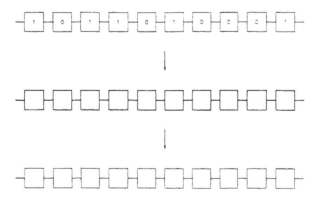

Fig. 9. Insertion of a "spacer" to decouple the functionalized chain ends

If the efficiency of the spacer can be modulated from the outside, perhaps by light absorption, we will already have a molecular logic device: There are two inputs - one light pulse with frequency ν_{exc} exciting one end, another with frequency ν_{mod} modulating the spacer - and there is one output: a light pulse with frequency ν_{probe} probing whether the excitation has arrived at the other end.

It should be stressed that switching in such a molecule will be a stochastic process. The switch does not change between "on" and "off", only the tunneling probability changes. To perform reliable calculations based on such elements the circuits will have to have a considerable amount of redundancy to compensate for the frequent switching "errors".

An extreme case of such redundancy and parallel computing would be a cellular automaton. This is an array of cells which can have at least two logical states and which are switched by clock pulses, whereby the switching probability depends on the state of the neighbour cells. A cellular automaton is schematically shown in Fig. 10.

Fig. 10. Scheme of a cellular automaton

For the sake of discussion and of formulating questions a model of a polyene-based cellular automaton is shown in Fig. 11. The cells are polyenes with donor and acceptor end groups, the states of the cells are those shown in Fig. 8 and the influence of the neighbour cells is via the dipole moment which detunes the frequency necessary to switch between States A and B. It is assumed that the excited state is metastable and lives "sufficiently long", an unrealistic assumption for molecules as simple as those of Fig. 8. It is fairly easy to estimate the error probability of such an automaton. In Fig. 12 the error probability is plotted as a function of the length of the clock pulses with the

Fig. 11. Model of a polyene-based cellular automaton

dipolar detuning as parameter, both in units of τ or $1/\tau$, where $1/\tau$ is the width of the resonance line for switching (Roth et al. 1989). According to these estimates the polyene automaton of Fig. 11 would perform very badly. The length of the clock pulse probably could be optimized, but the dipolar detuning will be much smaller than $1/\tau$!! So one would have to look for other ways of coupling between the cells and for more complex molecules, in which the resonance for switching is much sharper than in simple polyenes.

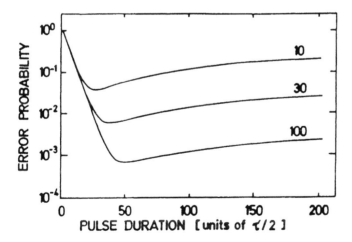

Fig. 12. Error probability of a cellular automaton like that in Fig. 11

4. CONCLUSION

This survey is by no means comprehensive. I have only sketched <u>one</u> of the proposal routes towards molecular electronics. But this sketch shows how many open questions there are, how difficult it is to formulate these questions in a relevant way, and how long it will take even to find out whether "small up" will be more promising than "large down". If we want to have molecular electronics in some decades we have to work now.

ACKNOWLEDGEMENT

This contribution is based on work funded by the ESPRIT Basic Research Actions "OLDS" and "MOLSWITCH" and by the Sonderforschungsbereich "Molekulare Elektronik". I want to thank my colleagues within these cooperations for many fruitful discussions.

REFERENCES

Aviram A 1989 *Molecular Electronics - Science and Technology* (New York: Engineering Foundation)

Barzoukas M, Blanchard-Desce M, Josse D, Lehn J M and Zyss J 1989 *Chem. Phys.* **133** 323

Bleier H, Lobentanzer H, Leising G and Roth S 1987 *Europhysics Letters* **4** 1397

Bleier H, Roth S, Shen Y Q, Schäfer-Siebert D and Leising G 1988 *Phys. Rev. B* **38** 6031

Carter F L 1982 *Molecular Electronic Devices* (New York and Basel: Marcel Dekker)

Chiang C K, Fincher C R, Park Y W, Heeger A J, Shirakawa H, Louis E J, Gau S C and MacDiarmid A G 1977 *Phys. Rev. Lett.* **39** 1098

Effenberger F, Schlosser H, Bäuerle P, Maier S, Port H and Wolf H C 1988 *Angewandte Chemie* **100** 274

Friend R H 1987 *Nature* **326** 335

Göhring W, Roth S and Hanack M 1987 *Synthetic Metals* **17** 259

Hanack M and Roth S 1990 *International Conference on Science and Technology of Synthetic Metals - ICSM '90* in Tübingen

Heeger A J, Kivelson S, Schrieffer J R and Su W P 1988 *Review of Modern Physics* **60** 781

Heeger A J 1989 *Faraday Discussions* **88** 203

Joachim C 1990 *The New Journal of Chemistry* in print

Kivelson S and Heeger A J 1988 *Synthetic Metals* **22** 371

Landauer R 1970 *Phil. Mag.* **21** 86

Li D, Minami N, Ratner M A, Ye C, Marks T J, Yang J and Wong G K 1989 *Synthetic Metals* **28** D585

Naarmann H and Theophilou N 1987 *Synthetic Metals* **22** 1

Ovchinnikov A A, Spector V N 1988 *Synthetic Metals* **27** B615

Pietronero L 1983 *Synthetic Metals* **8** 285

Roth S 1986 *Z. Physik B - Condensed Matter* **64** 25

Roth S and Bleier H 1987 *Advances in Physics* **36** 385

Roth S, Mahler G, Shen Y Q and Coter F 1989 *Synthetic Metals* **28** C815

Roth S and Filzmoser M 1990 *Advanced Materials* in print

Skotheim T A 1986 *Handbook of Conducting Polymers, Vols. I and II* (New York and Basel: Marcel Dekker)

Thoughless D J 1977 *Phys. Rev. Lett.* **39** 1167

Tsukamoto J, Takahashi A and Kawasaki K 1990 *Jap. J. of Appl. Phys.* **29** 125

Yacoby Y and Roth S 1985 *Solid State Communications* **56** 319

Yu Lu 1988 *Solitons and Polarons in Conducting Polymers* (Singapore: World Scientific)

The controlled electromagnetic response of polyanilines and its application to technologies

Arthur J. Epstein
Department of Physics and Department of Chemistry, The Ohio State University, Columbus, Ohio 43210-1106 USA

Alan G. MacDiarmid
Department of Chemistry, University of Pennsylvania, Philadelphia, PA 19104-6323

ABSTRACT

It is attractive to consider the polyanilines for applications in a wide variety of technologies because of their flexible chemistry, solubility, processibility, environmental stability, and ease of forming into blends and composites. We review here the response of the polyanilines to electric fields at frequencies ranging from dc through audio and microwave frequency to the optical frequency range. Potential applications of the polyanilines in commodity, commodity/electrochemical, and "high tech" applications for all three ranges are discussed. In addition, potential applications of the polyanilines based upon their chemical and electrochemical responses are pointed out.

INTRODUCTION

Conducting polymers have been under active study since the report in 1977 [1] that polyacetylene can be p-and n-doped to high conductivity. Extensive progress has been made in understanding the underlying chemistry and physics of the conducting polymers and to broadening the study to many other polymer systems, including polypyrroles, polythiophene, and polyphenylenevinylenes [2, 3]. However, applications have been slow in commercialization, in part due to issues of environmental stability, solubility, processibility, compatibility in blends and composites, as well as cost. Although the polyanilines have been known for over a century [4-6], their intensive study dates only to the mid-1980's [7-13]. With an increase in the understanding of the chemistry and processibility of polyanilines their feasibility for application in a variety of technologies can now be examined.

Though frequently termed "conducting polymers", the potential applications of the electronic polymers span a very large range of conductivities (from insulating through "metallic") and a broad range of frequency response (from dc through optical frequencies). Fig. 1 schematically summarizes this perspective. The potential types of applications are broken down into three differing types: COMMODITY, COMMODITY/ELECTROCHEMICAL, and "HIGH TECH".

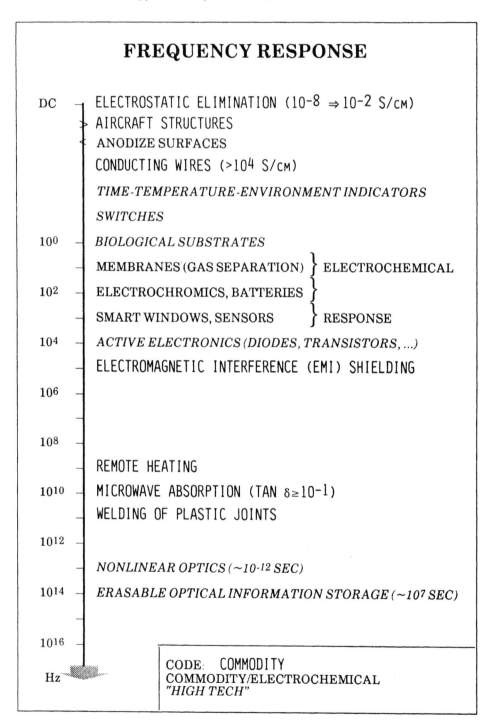

Figure 1

COMMODITY applications of conducting polymers generally use the polymers' bulk conductivity to achieve some commercially important function. Examples would be for electrostatic elimination (e.g., in carpets or surface coatings) and in electromagnetic interference (EMI) shielding, in addition to remote heating and microwave absorption. It is noted that there are differing electromagnetic frequencies involved and that the requirements for conductivity span a very wide range from as low as 10^{-8} S/cm for electrostatic elimination to values greater than 10^4 S/cm for commercially useful lightweight conducting wires.

The second class of applications, COMMODITY/ELECTROCHEMICAL, utilizes the electrochemical response of these polymers to applied potential [8-10, 13, 14]. Examples of technologies where this is important are batteries, electrochromics, and smart windows. The latter concept requires the conducting polymer to change some property, for example, optical transmission or conductivity in response to a stimulus, for example, electrochemical potential or light. The commodity and commodity/electrochemical applications will require the inexpensive production of a substantial volume of material to be commercially successful.

The *"HIGH TECH"* applications utilize either a particularly high performance characteristic of the polymers or an unusual combination of properties that are not readily available in other materials. Among these properties are the combined response to environment and temperature to yield time-temperature-environment indicators. Examples here include materials whose conductivity changes with time, or whose color or conductivity changes when exposed to chemicals such as ammonia. These applications usually occur on time scales of milliseconds or longer. Other high tech applications include active electronic elements such as diodes and transistors, although study of current state-of-the-art of electronic polymers shows their performance is substantially lower than that of inorganic semiconductors based on silicon and GaAs. Novel *"HIGH TECH"* applications include utilization of electronic polymers as the active elements in nonlinear optics and also for optical information storage. These applications involve the frequency in which the response is in the optical regime ($>10^{12}$ Hz) with nonlinear optics requiring fast response ($\sim 10^{-12}$ sec) and erasable optical storage involving long time stability ($\sim 10^7$ sec).

CHEMISTRY AND PROCESSING OF POLYANILINES

The chemistry of the polyanilines has been recently reviewed [14, 15]. The oxidation state of the insulating base forms of the polymer can be varied from that of fully reduced leucoemeraldine (LEB), to the half-oxidized emeraldine base, and the fully oxidized pernigraniline base (PNB), Fig. 2. The insulating base forms (conductivity, $\sigma \sim 10^{-10}$ S/cm [16]) can be converted to the conducting form ($\sigma \sim 10^0 - 4 \times 10^2$ S/cm) by equilibration with protonating acids to form the emeraldine salt, Fig. 2. It was shown that there is a dramatic change in electronic structure in going from the insulating to the conducting form [17].

One of the most attractive aspects of the polyanilines is the ability to derivatize the polymer at the ring or nitrogen sites to improve their processibility and compatibility. In particular two derivatives are pointed out here, the methyl ring derivative, poly(orthotoluidine) which replaces a single hydrogen of each C_6 ring with single CH_3 unit [18] and the sulfonic acid ring substituted derivative [19] which

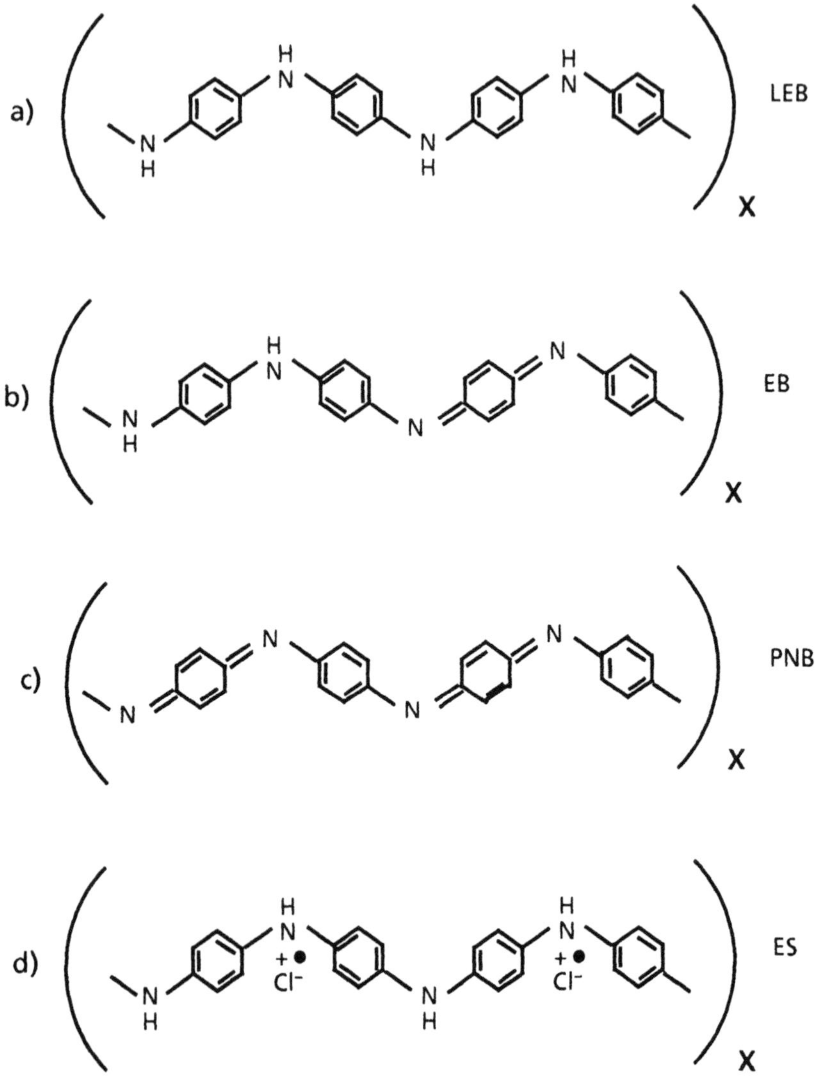

Fig. 2 Oxidation states of polyaniline characterized by the fraction of oxidized (1-y) and reduced (y) repeat units: (a) leucoemeraldine base (1-y=0); (b) emeraldine base (1-y=0.5); (c) pernigraniline base (1-y=1); and (d) emeraldine salt polymer.

replaces a hydrogen on every other ring with an $SO_3^-H^+$ unit, Fig. 3. The former system has been particularly important in enabling critical study of the effects of ring derivatization and interchain separation on electronic properties [20] while the latter system introduces new opportunities in water solubility and solubility in aqueous base media. The parent polyaniline in the emeraldine base form is soluble in *N*-methylpyrrolidinone (NMP) from which it can be cast to give lustrous strong films [21]. Polyaniline is also soluble in acetic acid [21] and sulfuric acid [4, 22]. Hence organic and aqueous solvents can be used, as well as basic, neutral or acidic media.

It has recently been shown that polymer films cast from NMP can in turn be monoaxially and biaxially oriented [23] by applying stress and simultaneously heating the films above the glass transition temperature [24]. The initially prepared unoriented films have a tensile strength of 55 MPa increasing to ~125 MPa for four-fold stretched films [23] overlapping the values [25] of Nylon 6 and Nylon 66. It is also recently reported that strong fibers of polyaniline can be drawn from emeraldine base [26] and salt [22] solutions. The ability to incorporate polyaniline in poly(paraphenylendiamine)-terephthalic acid, the polymer for Kevlar® aramid, has also been demonstrated [22,27]. There has been substantial progress recently in elucidating the local crystal structure in polyanilines [28, 29]. The polymers as currently prepared generally are up to 50 percent crystalline. Extensive studies have shown that the electronic phenomena in polyanilines are strongly dependent upon the conformations of the C_6 rings [30].

LOW FREQUENCY APPLICATIONS

For applications that require dc through audio frequency ($\sim 10^4$ Hz) the charge carriers (that is, electrons or holes) must essentially cross the entire sample. This requires continuity of the conduction path for the electrons and holes through the material. Applications in this area range from antistatics in carpets and clothing to forming conducting matrices for graphite fiber re-enforced plastics. One area of promise is as part of plastic composites for aircraft structures. In this application it is important that the matrix for graphite fibers be conductive to avoid damage to the aircraft upon lightening strikes. It is noted that the current commercial alternatives in this area involve largely carbon black or metal flake loaded insulating polymers [31]. Though readily available commercially, these loaded polymers generally are not easily tuned for conductivities in the range of 10^{-6} S/cm through 10^{+1} S/cm due to the steepness of the percolation threshold. Samples in the higher conductivity range often have degraded mechanical properties because of the high loading of the conducting particulates. A further problem in the existing composite polymer technologies is aging at particle-particle interfaces leading to changes in the conductivity with time.

The broad range of conductivities as a function of substituents and temperature enables customizing the polyanilines to particular applications. Figure 4 shows the conductivity vs. $T^{-\frac{1}{4}}$ for a four-fold stretched polyemeraldine film that has been doped to $x \equiv [Cl]/[N] = 0.5$ [32], while Figure 5 displays the pH behavior of the conductivity of sulfonated polyaniline [19, 33]. This material has a pH independent conductivity in the range of pH ≤ 7.5, spanning the range of biologically important pH values. The temperature dependence of the conductivity of sulfonated polyaniline and its thermoelectric power are shown in Fig. 6. The poly(orthotoluidine) emeraldine

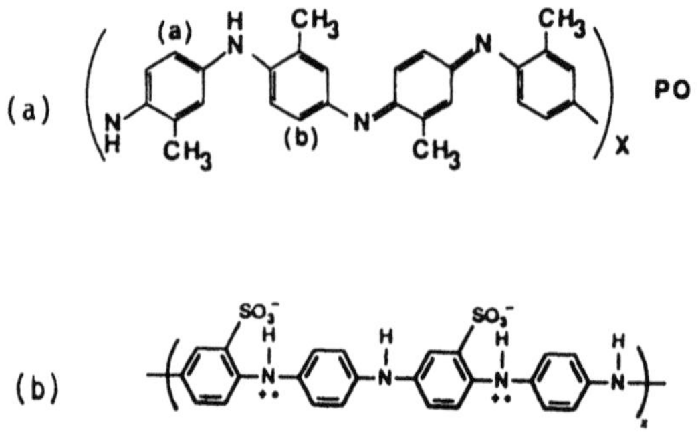

Fig. 3(a) Poly(orthotoluidine)
(emeraldine base form); (b) sulfonated
polyaniline (self-doped salt form).

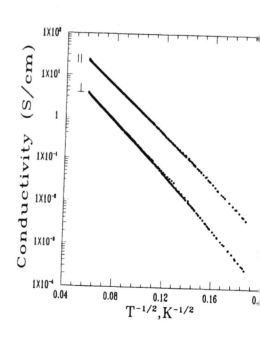

Fig. 4 Temperature dependence of dc
conductivity of 50% doped 1:4 stretched
PAN-ES (HCl) films in the directions
parallel and perpendicular to the
stretching directions (from Ref. 32).

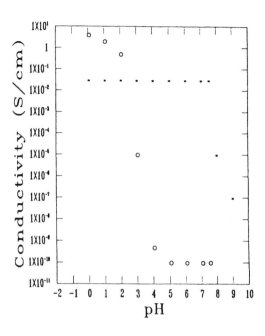

Fig. 5 pH dependence of conductivities for sulfonated polyaniline (∗) and the meraldine form of polyaniline (o) (from ref. 33).

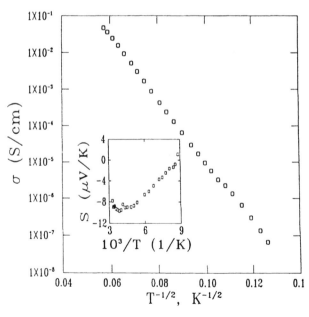

Fig. 6 Temperature dependence of the conductivity of sulfonated polyaniline. The inset shows the thermopower vs. reciprocal temperature for sulfonated polyaniline (from Ref. 33).

hydrochloride salt has a room temperature conductivity of $\sim 10^{-2}$ S/cm and a temperature dependence similar to that of the sulfonated polyaniline [20].

MICROWAVE FREQUENCY RESPONSE

As the frequency increases to the radiowave-microwave-millimeter wave range (10^7-10^{11} Hz) the charge carriers need not cross the entire sample in order to contribute to the conductivity. Therefore regions of higher conductivity that are not part of a connected path to the electrodes can contribute to the response. Potential applications for conducting polymers include in the area of microwave (radar) absorbers, screening of electromagnetic interference and microwave antennae. For efficient use of absorbers without any significant reflectance, it is important to grade the microwave absorption with thickness. The absorption is directly related to the quantity

$$\tan \delta = \sigma / (\omega \varepsilon \varepsilon_0) = \varepsilon'' / \varepsilon' \qquad \qquad \textit{Equation 1}$$

where $\tan\delta$ is the loss tangent, σ and ε the conductivity and dielectric constant at frequency ω. Here ε_0 is the dielectric constant of vacuum and ε' and ε'' the real and imaginary parts of the dielectric constant. Because of the ability of polyanilines to absorb microwave and electromagnetic radiation, unusual opportunities are open for areas involving remote heating of materials and surfaces. This occurs when the conducting polymer absorbs a significant fraction of the radiation impinging upon it. One application of this remote heating is the "welding" of plastic joints. Conducting polyaniline may be placed between two pieces of plastic (for example, high density polyethylene) whose surface needs to be melted in order to fuse them together. Exposing two components of insulating polymer with a layer of polyaniline between them to microwave radiation can lead to the formation of a strong bond. A related application of remote heating is to form part of a polymer vessel of a low melting polymer which has the conducting polyaniline applied to it. Then exposure of this configuration to the microwave radiation will lead to melting of the polymer adjacent to the polyaniline and hence forming a remote break-seal.

There is a wide range of behaviors available in polyaniline polymers depending upon the choice of orientation, substituents. and protonation level as well as counterion. For example, Figure 7 summarizes the temperature dependence of the microwave tan δ for four-fold stretch oriented emeraldine hydrochloride, poly(orthotoluidine) hydrochloride and sulfonated polyaniline. It is noted that even at room temperature, the tanδ of emeraldine salt spans an order of magnitude depending on orientation and substituents. Graded absorbers can be prepared by varying the protonation level with polymer thickness. Smaller tanδ's can be obtained using lower doping levels [34]. Currently, carbon black loaded insulating polymers are used to absorb the electric field component of electromagnetic radiation. Alternatively, ferrite loaded insulating polymers are used to absorb the magnetic field component of electromagnetic radiation [35].

OPTICAL FREQUENCY RESPONSE

Electronic polymers such as polyacetylene, polydiacetylene, and related materials have been examined for their nonlinear optical susceptibility [36]. Though the

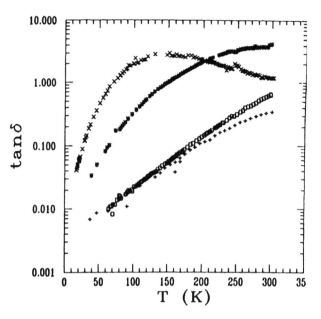

7 Tan δ at 6.5 GHz vs. temperature four-folded stretched emeraldine rochloride parallel to the stretch ection (x), perpendicular to the etch direction (∗), unoriented ν(orthotoluidine) hydrochloride (o) unoriented self-doped sulfonated raniline (+) (see Refs. 20, 32, and 33).

Table I. Materials exhibiting long-lived photoinduced effects.

Material	Temperature (K)	Lifetime
Emeraldine base film	77	110 min
Emeraldine base in KBr	77	24 hr
Pernigraniline base in KBr	77	≫24 hr
Pernigraniline base in KBr	200	>24 hr
Emeraldine poly(o-toluidine) in KBr	77	≫24 hr
Emeraldine poly(o-toluidine) in KBr	250	∼2.5 hr
Pernigraniline poly(o-toluidine) in KBr	77	≫24 hr
Pernigraniline poly(o-toluidine) in KBr	250	≫24 hr
Pernigraniline poly(2-ethoxyaniline) in KBr	77	140 min

insulating forms of polyaniline have substantial $\chi^{(3)}$ values [37] their relatively high optical absorbances [38, 39] prevent their serious consideration as practical $\chi^{(3)}$ materials at this time.

A novel application of the polyaniline family of polymers relies upon the existence of long-lived optically active photoexcited defects in the polymers [40]. This feature allows for a possible new method of storage of optical information [41] as an alternative to magneto-optical effects in rare earth-transition metal alloys [42] and amorphous to crystalline phase transitions in chalcogenide glasses [43]. Photoexcitation of several forms of polyaniline result in long-lived changes in the absorption spectra [38, 40, 41] associated with a charged defect trap state [30, 38]. These new absorptions exhibit long-lifetimes at low temperatures (below~250K), with erasure occurring upon warming the sample. Table I summarizes some of the initial results for polyaniline and its derivatives. The Table lists the material, the oxidation state, the temperature at which the measurement was carried out and the lifetime for the photoinduced images. In each case the photoexcitation was carried out using a argon ion laser or a mercury arc lamp and probed at ~1.5 eV. Studies of other derivatives suggests that there are polyanilines whose capacity for optical information storage is stable at room temperature, with erasure above room temperature or possibly through selected exposure of individual stored bits of information to laser light.

SUMMARY

The polyaniline family of polymers is a flexible system that enables tuning of polymer properties over very broad chemical, electrochemical and physical ranges. Potential commodity, commodity/electrochemical and high tech applications of conducting polymers exist in the frequency range from dc through to optical frequencies. Continued development of these materials and their processibility may lead to the applications of these systems.

ACKNOWLEDGMENT

This work was supported in part by the Defense Advanced Research Projects Agency through a contract monitored by the U.S. Office of Naval Research. The authors acknowledge the stimulating interaction with their students, postdoctoral fellows, and associates.

REFERENCES

1.   Shirakawa H, Louis E J, MacDiarmid A G, Chiang C K and Heeger A J 1977 J. Chem. Soc. Chem. Commun. 578
2.   See, for example, 1986 Skotheim T (ed.), Handbook of Conducting Polymers Vols 1 and 2, Dekker, New York
3.   See, for example, Proceedings of the International Conferences on the Science

and Technology of Synthetic Metals, Tübingen, Germany, September 1990 [1991 Synth. Met. in press]; Santa Fe, NM, June 1988 [1988 Synth. Met. **27**; 1989 **28 29**]; Kyota, Japan, June 1986 [1987 Synth. Met. **17-19**]; Albano Terme, Italy, June 1984, [1985 Mol. Cryst. Liq. Cryst. **117**].

4. Green A G and Woodhead A E 1910 J. Chem. Soc. Trans. **97** 2388; 1912 **101** 1117
5. Willstatter R and Cramer C 1910 Chem. Ber. **43** 2976; 1911 **44** 2162; Willstatter R and Dorogi S 1911 **44** 2166
6. DeSurville R, Jozefowicz M, Yu L T, Perichon J and Buvet R 1968 Electrochem. Acta **13** 1451
7. Humphrey, B D, Chiang J C, Huang W S, Richter A J, Somasiri N L D, MacDiarmid A G, Yang X Q, Epstein A J and Bigelow R W 1985 Bull. Am. Phys. Soc. **30** 605
8. MacDiarmid A G, Chiang J-C, Halpern M, Huang W-S, Krawczyk J R, Mammone R J, Mu S L, Somasiri N L D and Wu W 1984 Polym. Prepr. **25** 248; MacDiarmid A G, Chiang J-C, Halpern M, Huang W-S, Mu S L, Somasiri N L D, Wu W and Yaniger S I 1985 Mol. Cryst. Liq. Cryst. **121** 173; Chiang J C and MacDiarmid A G 1986 Synth. Met. **13** 193
9. Genies E M, Syed, A A, and Tsintavis C C 1985 Mol. Cryst. Liq. Cryst. **121** 181
10. Paul E W, Ricco A J and Wrighton M S 1985 J. Phys. Chem. **89** 1441
11. Travers J P, Chroboczek J, Devreux F, Genoud F, Nechtschein M, Syed A, Genies E M and Tsintavis C 1985 Mol. Crsyt. Liq. Cryst. **121** 195
12. Salaneck W R, Lundstrom I, Huang W S and MacDiarmid A G 1986 Synth. Met. **13** 291
13. McManus P M, Yang S C and Cushman R J 1985 J. Chem. Soc. **1556**
14. MacDiarmid A G and Epstein A J 1989 Faraday Discuss. Chem Soc. **88** 317
15. MacDiarmid A G and Epstein A J 1990 Mat. Res. Soc. Symp. Proc. **173** 283
16. Zuo F, Angelopoulos M, MacDiarmid A G and Epstein A J 1989 Phys. Rev. B **39** 3570
17. Epstein A J, Ginder J M, Zuo F, Bigelow R W, Woo H S, Tanner D B, Richter A F, Huang W S and MacDiarmid A G 1987 Synth Met. **18** 303
18. Wei Y, Focke W W, Wnek G E, Ray A and MacDiarmid A G 1989 J. Phys. Chem. **93** 495; Ray A, MacDiarmid A G, Ginder J M and Epstein A J 1990 Proc. Mat. Res. Soc. **173** 353
19. Yue J and Epstein A J 1990 J. Am. Chem. Soc. **112** 2800
20. Wang Z H, Javadi H H S, Ray A, MacDiarmid A G and Epstein A J 1990 Phys. Rev. B **42** 5411
21. Angelopoulos M, Asturias G E, Ermer S P, Ray A, Scherr E M, MacDiarmid A G, Akhtar M, Kiss Z and Epstein A J 1988 Mol.Cryst. Liq. Cryst. **160** 151
22. Andreatta A, Cao Y, Chiang J C, Heeger A J and Smith P 1988 Synth. Met. **26** 383; Andreatta A, Tokito S, Smith P and Heeger A J 1990 Mat. Res. Soc. Symp. Proc. **173** 269
23. Cromack K R, Jozefowicz M E, Ginder J M, McCall R P, Epstein A J, Scherr E and MacDiarmid A G 1989 Bull. Am. Phys. Soc. **34** 583; Cromack K R, Jozefowicz M E, Ginder J M, McCall R P, Du G, Kim K, Li C, Wang Z H, Epstein A J, Druy M A, Glatkowski P J, Scherr E M and MacDiarmid A G submitted
24. Wei Y, Jang G W, Hsueh K F, Scherr E M, MacDiarmid A G and Epstein A J 1989 Polymeric Materials Science and Engineering **61** 916; and to be published
25. Handbook of Chemistry and Physics, 1977 57 edited by R C West CRC Press Cleveland

26. Tang X, Scherr E, MacDiarmid A G and Epstein A J 1989 Bull. Am. Phys. Soc. **34** 583; to be published
27. Hsu C H, Vaca-Segonds P and Epstein A J 1991 Synth. Met. in press
28. Jozefowicz M E, Laversanne R, Javadi H H S, Epstein A J, Pouget J P, Tang X and MacDiarmid A G 1989 Phys. Rev. B **39** 12,958
29. Pouget J P, Jozefowicz M E, Epstein A J, Tang X and MacDiarmid A G 1990 Macromolecules **23** xxx
30. Ginder J M and Epstein A J 1990 Phys. Rev B **41** 10674
31. Sichel E Carbon Black-Polymer Composites, The Physics of Electrically Conducting Composites 1982 Marcel Dekker, Inc., New York and Basel
32. Wang Z H, Scherr, E, MacDiarmid A G and Epstein A J submitted
33. Yue J, Wang Z H, Cromack K R, Epstein A J and MacDiarmid A G submitted
34. Javadi H H S, Cromack K R, MacDiarmd A G and Epstein A J 1989 Phys. Rev. B **39** 3579
35. Emerson and Cuming 869 Washington Street, Canton, Mass. 02021 "Microwave Absorbers" Catalog
36. See, for example, Proc. Symp. on Advanced Organic Materials, 1990 Mat. Res. Soc. Symp. Proc. 173 ed by Chiang L Y, Chaiken P M, and Cowan D D
37. Ginder J M, Epstein A J and MacDiarmid A G 1989 Synth. Met. **29** E395
38. McCall R P, Ginder J M, Leng J M, Ye H J, Manohar S K, Master J G, Asturias G E, MacDiarmid A G and Epstein A J 1990 Phys. Rev. B **41** 5202
39. Leng J M, Ginder J M, McCall, R P, Ye H J, Epstein A J,. Sun Y, Manohar S K and MacDiarmid A G 1991 Synth. Met. in press
40. McCall R P, Ginder J M, Roe M G, Asturias G E, Scherr E M, MacDiarmid A G and Epstein A J 1989 Phys. Rev. B **39** 10,174
41. McCall R P, Ginder J M and Epstein A J submitted
42. See, for example, numerous articles and references in 1982, 1983, 1984, 1985, 1988 Proc. Soc. Phot. Opt. Inst. Eng. **329, 382, 420, 490, 529, 899**
43. Huijser A 1984 Physica B **127** 90

Attempts to synthesize low gap aromatic polymers

M. Hanack[*], G. Hieber, G. Dewald, H. Ritter

Institut für Organische Chemie, Lehrstuhl für Organische Chemie II der Universität Tübingen, Auf der Morgenstelle 18, D-7400 Tübingen, Federal Republic of Germany

1. Introduction

Polyarenemethylidenes (PAM) have been predicted to be low gap polymers with good semiconducting and photoelectrical properties (Brédas 1985). Based on VEH (Valence Effective Hamiltonian) calculations these systems should have a bandgap E_g of about 1 eV a value also observed in the case of polyisothianaphthene (PITN) (Wudl 1984).

An important prerequisite to obtain low gap polymers is, according to the theoretical work of Brédas (1987), some quinonoid contribution in the groundstate of the polymer, an electronic property also present in PITN.

The necessity of quinonoid contributions in the polymer backbone prompted us to study methinebridged aromatic polymers which contain alternating aromatic and quinonoid subunits. A polymer of this type is shown in the general formula I.

Several approaches to synthesize this type of polymers I based on benzene and thiophene subunits have been reported in the literature (Jenekhe 1986, Giles 1987). In most of the experiments however, no information has been given about the detailed structure of the polymer backbone. Attempts to repeat the described procedures met with no success by us as well as by others (Jira 1987a, Wudl 1988).

Another approach is the condensation of heterocyclic aldehydes catalyzed by Lewis acids, which results in the formation of polymers with conductivities of 10^{-4} S/cm, which were described to have a composition given in formula I (Bräunling 1989, Jira 1987b).

We decided to follow a different approach for the synthesis of polymers or oligomers containing aromatic and quinonoid structures in the backbone. Contrary to all the attempts which have been described, we developed a systematic route to synthesize this type of polymers I starting with very well defined monomers. The requirement for this purpose is the preparation of suitable monomeric precursors in which beside the aromatic also the quinonoid structure is already present.

general formula :

polyarenemethylidene

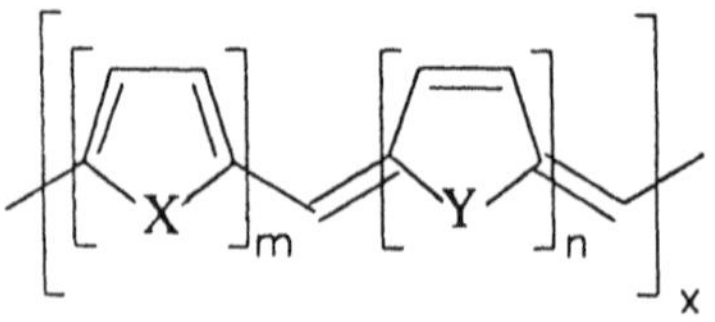
X
m
Y
n
x

I

(X, Y = SO_2, SO, S, N-R, CH=CH)

Scheme 1 shows the principle routes for the synthesis of polyarenemethylidenes (PAM) using chemical and electrochemical methods. The different polymerization methods require different precursors as starting materials.

Scheme 1

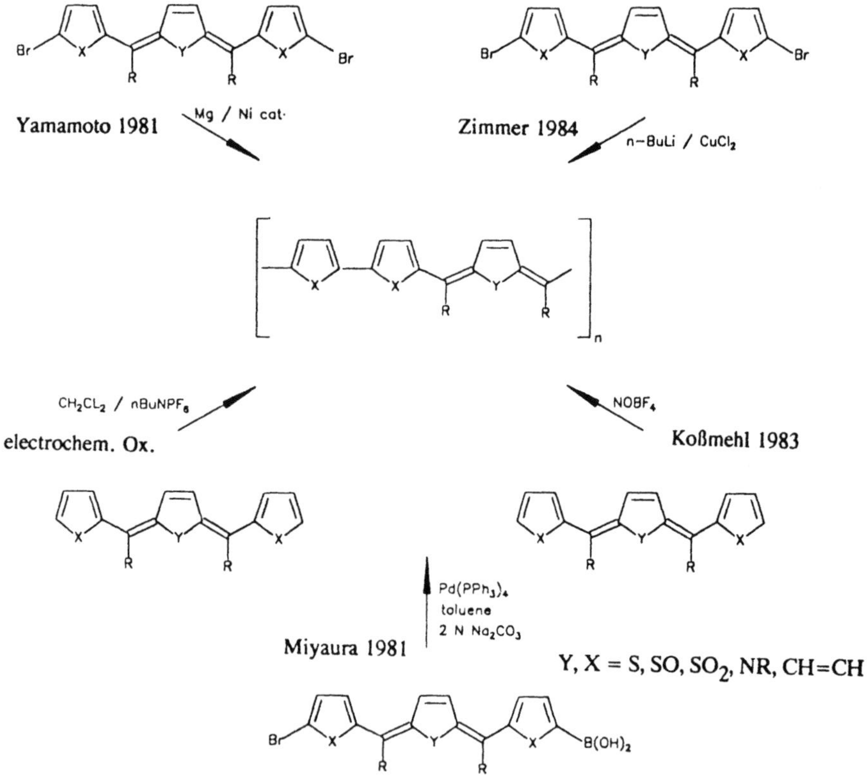
Br
Br
Yamamoto 1981
Mg / Ni cat.
Zimmer 1984
n–BuLi / $CuCl_2$
n
CH_2CL_2 / $nBuNPF_6$
electrochem. Ox.
$NOBF_4$
Koßmehl 1983
$Pd(PPh_3)_4$
toluene
2 N Na_2CO_3
Miyaura 1981
Y, X = S, SO, SO_2, NR, CH=CH
$B(OH)_2$

In the following we describe the synthesis of monomeric compounds as suitable precursors for polymerisation according to scheme 1.

2. Benzenoid and anthracenoid precursors

The specific linkage of monomers with benzenoid units in the chemical polymerization re-action affords the presence of reactive substituentes like bromine in the p-positions of the peripheric rings.

A suitable precursor is the tetraphenylsubstituted quinodimethane 3. This airsensitive mo-nomer is obtained by treating the diole 1 with hydrobromic acid in acetic acid and subse-quent reduction of the resulting dibromide 2 with zincdust.

Ring annellation of the quinonoid ring leads to more stable anthracenoid derivates (Hanack 1989).Wittig olefination of anthraquinone in THF afforded 9,10-dihydro-9,10-bis-((α-4'-bromphenyl)-α-phenylmethylidene)-anthracene 4 , an airstable yellow compound slightly soluble in polar solvents.

The monomers 3,4 can be easily converted into metalorganic intermediates. The addition of transition metal catalyst starts cross coupling reaction yielding the corresponding poly-mers (s. scheme 1)(Zimmer 1984, Yamamoto 1981).

As shown in the preparation of poly-p-phenylene PPP (Miyaura 1981) an alternative polymerization route can be used starting from the boronic acid 5 with Pd(PPh$_3$)$_4$ catalyst (s. scheme 1).

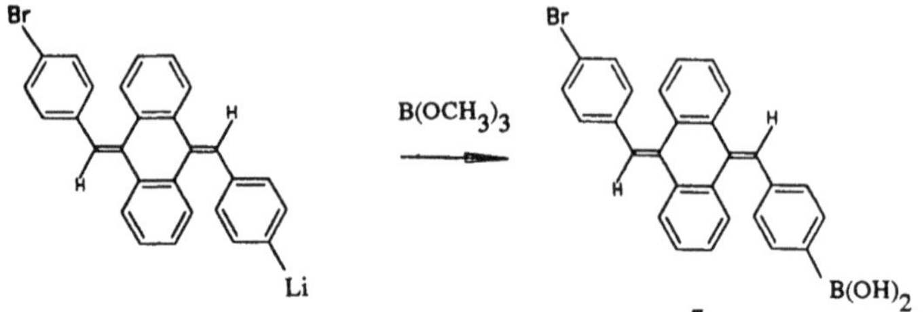

5

A Wittig olefination as described above can be used in a similar way for higher oligomers like 1,4-bis-(9',10'-dihydro-10'-phenylmethylidene-anthr-9'-ylidenemethyl)-benzene 6.

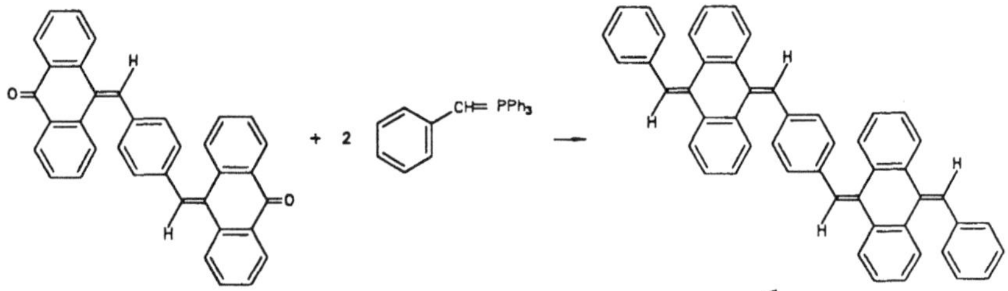

6

3. Heteroaromatic precursors

The unsubstituted heteroaromatic rings are more reactive with respect to chemical and electrochemical polymerization.

In a Knoevenagel type condensation of heteroaromatic aldehydes with 3-sulfolene or its benzocondensed analogues (Hanack 1990, Ritter 1990) we obtain a serie of methin-bridged airstable oligomers 7, 8, 9 and 10. The synthesis of brominated compounds like 7 by using the brominated aldehyde as educt is also possible.

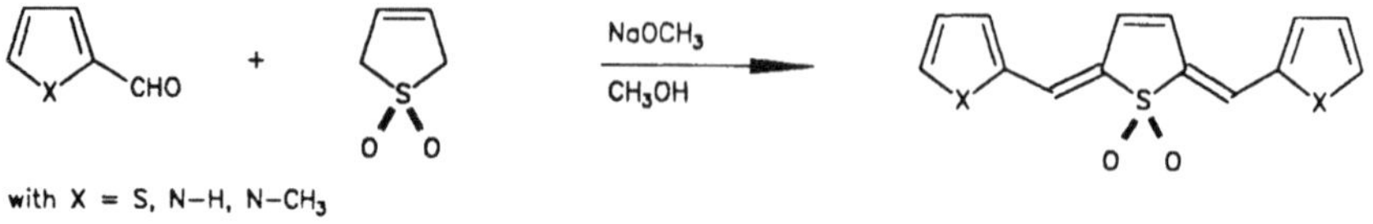

with X = S, N−H, N−CH$_3$

8

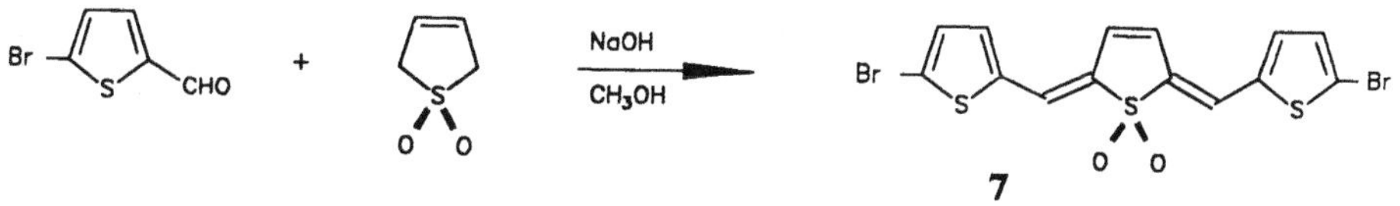

7

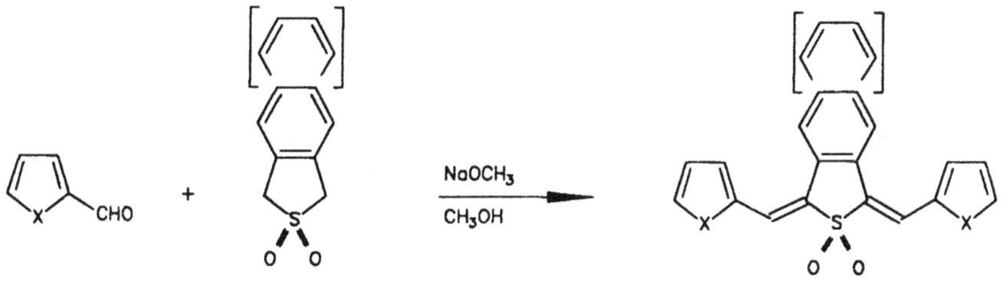

with X = S, N–H, N–CH₃

9 **[10]**

A crystal structure of 2,5-bis-(2-thienylmethylidene)-2,5-dihydrothiophene-1,1-dioxide **8** exhibits planar geometry and reduction (elongation) of the single (double) bonds indicating π-conjugation (Hanack 1990). The molecule crystalizes in a monoclinic lattice in the space group $P2_1/c$ (R=3.4%) (lattice constants : a=1522.7 pm b=1225.8 pm c=730.5 pm $\alpha=\gamma=90°$ $\beta=80.39°$).

Because of the inertia of the sulfone-group against reduction we synthesized the corresponding sulfoxides 11, 12 and 13 by condensation of heteroaromatic aldehydes with 2,5-dihydrothiophene-1-oxide and its benzocondensed analogues in basic media (Hanack 1990, Ritter 1990).

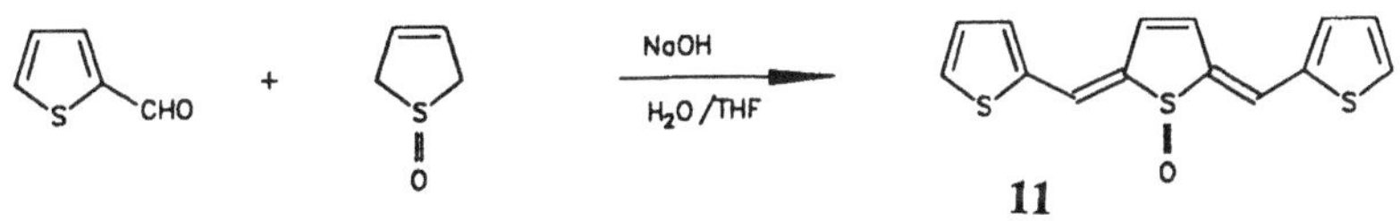

11

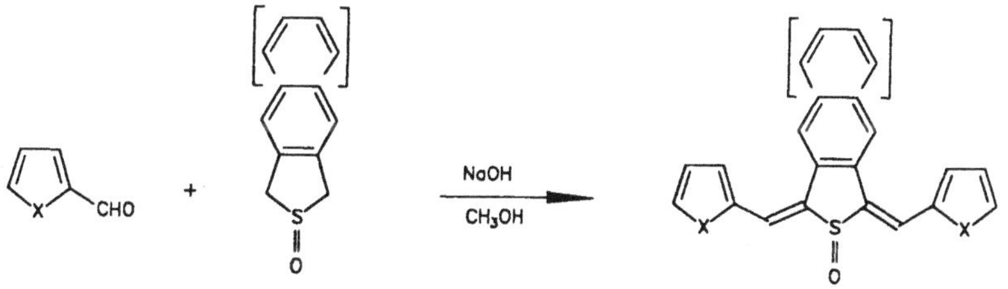

mit X = S, N–CH₃

12 **[13]**

The sulfoxide group can be easily reduced using 2-chloro-1,3,2-benzodioxaphosphole (Chasar 1976) in presence of pyridine yielding airsensitive 2,5-bis-(2-thienylmethylidene)-2,5-dihydrothiophene 14 and its more stable benzocondensed analogues 15 and 16 (Hanack 1990, Ritter 1990).

For the synthesis of methin-bridged oligomers with a sterical high demand another synthetic route has been developed.

Treatment of the diole 17 in toluene with aqueous hydroiodic acid under rapid stirring affords directly the 2,5-bis-[di-(2-thienyl)methylidene]-2,5-dihydrothiophene 18 under formation of iodine which is trapped with sodiumdithionite ($Na_2S_2O_4$).

X-ray analysis exhibits the molecule in a non planar arrangement. Both peripheric rings are twisted toward the inner quinonoid ring plane. The molecule crystalizes in a monoclinic lattice in the space group $P2_1/n$.(lattice constants : a=1234.0 pm b=1256.8 pm c=1359.1 pm $\alpha=\gamma=90°$ ß=104.93°) (Hanack 1990).

The exact solution of the crystal structure is not possible because the peripheric thiophene rings are able to rotate between two energetically equivalent positions.

For polymerization of the heterocyclic monomers the direct oxidation route can be used because of the higher reactivity of the electron-rich heterocyclic precursors compared to the benzenoid derivates.

Several kinds of oxidants like $NOBF_4$, $NOSbF_6$, $FeCl_3$, $MoCl_5$, etc. (Koßmehl 1983) are possible for oxidative polymerisation yielding directly the polymer in its doped state.

For the dibromides e.g. 7 the cross coupling using metalorganic intermediates is also an interesting method affording the undoped polymers.

Electrochemical investigations with cyclic voltammetry indicates, that the oxidation of the monomers containing pyrrole rings occures at low oxidation potential. The sulfides 14, 15 and 16 are easier to oxidize than the corresponding oxygen containing sulfones 8, 9 and 10 and sulfoxides 11, 12 and 13.

First attempts of electrooxidative coupling of 14 and 18 yield in polymeric films with promising UV spectroscopic characteristics.

4. Conclusion

In this paper we present a general synthesis of methine-bridged oligomers in a great structural variety as precursors for conducting polymers. The chemical and elctrochemical polymerization is in progress and will lead to low gap polymers with a well defined structure.

5. References

Brédas J.L. 1985 Springer Ser. Sol. State Sci. **63** 166

Brédas J.L. 1987 Synth. Met. **17** 115

Bräunling H. Becker R. Blöchl G. 1989 Springer Ser. Sol. State Sci. **91** 465

Chasar D. W. Pratt T. M. 1976 Synthesis 262

Giles J. 1987 PCT Int. Appl. WO 87/00678

Hanack M. Dewald G. 1989 Synth. Met. **33** 409

Hanack M. Hieber G. Wurst K. Strähle J. 1990 Chem. Ber. to be submitted

Jenekhe S. A. 1986 Nature **322** 345

Jira R. Bräunling H. 1987a Synth. Met. **20** 375

Jira R. Bräunling H. 1987b Synth. Met. **17** 691

Koßmehl G. 1983 Makromolekulare Chemie Rapid Commun. 639

Miyaura N. Suzuki A. Yanagi T. 1981 Synth. Commun. **11** 513

Wudl F. Heeger A. J. Kobayashi M. 1984 J. Org. Chem. **49** 3382

Wudl F. Patil A. O. 1988 Macromolecules **21** 540

Yamamoto T. Sanechika K. Yamamoto A. 1981 Chem. Lett. 1079

Zimmer H. Amer A. Mulligan K. J. Mark H. B. Pons S. Mac Aleer J. F. 1984 J. Polym. Sci. Polym. Lett. Ed. **22** 77

Summary of the workshop: Perspectives for the 1990s

Professor D T Clark

Imperial Chemical Industries plc, PO Box 90, Wilton, Middlesbrough, Cleveland, TS6 8JE

1   INTRODUCTION

The field of Conducting Polymers spanning as it does interests in novel physics, novel polymer synthesis and processing and unique intrinsic properties is a rapidly moving one and perhaps a good place to start in any overview is to briefly consider how the balance of the themes/topics published in 1989 (as a reflection of the current research activity) and of the papers in this book (as a signpost for the future) compare. Figure 1 shows an analysis based on Dr Österholm's[1] reading of the 1989 literature and my own assessment of this particular meeting. The dominant change, as is readily apparent from these data, is the shift from assessment of properties towards assessment of potential applications with the concomitant increasing interest in processing.

RESEARCH FOCUS COMPARISONS

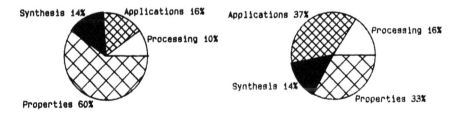

Figure 1

Conducting polymers fit into the classification of "effect" polymers and part of the current excitement arises from the fact that there are unique opportunities in this generic area to engineer combinations of exceptional properties spanning physical, chemical, mechanical, electrical/electronic, magnetic, electromagnetic and 4th state properties and many examples of the genre have been detailed in the preceding chapters in this book.

In general the applications that are already apparent or are envisaged imply a high degree of sophistication and low weight of material per unit product characteristic of effect materials and although in specific cases costs will inevitably reflect the expense of synthesis (the most notable exception to this is polyaniline) this will not preclude the relatively rapid adoption of these materials in end applications where the unique combinations of properties provides a cost effective total engineered solution. As in other speciality materials areas it is all too easy to get into the mindset that cost of material is everything, whereas in reality an expensive material with superior specific performance particularly if there are processing, design, fabrication advantages, more often than not can provide a more cost effective final engineered solution.

The particular features of the conducting polymer systems of the ability to "engineer" combinations of exceptional properties (modulus, conductivity, non linear optical etc) point to the need to reconceptualise the final engineered solution with this is mind rather than focusing on direct displacement by conducting polymers of conventional materials which inevitably places the focus on cost of materials rather than the total solution to a problem or opportunity. The preceding chapters perhaps under emphasise this point and the focus of research activity and debate about applications needs to be moved on from "replacement of copper" with a commodity mindset to the plane of, "levels of integration of function" which will be a feature of the generation of smart materials of the future and which will provide a genuine mimic of the living world. This does not detract from raising the level of awareness generally of new conducting polymer systems arising from early introductions into "commodity" applications such as the Bridgestone battery[2] or the conducting brushes in Xerox machines[3] and this has it's parallels in the ceramics industry where awareness of the capabilities of high performance ceramics is being raised in the general public's eyes by ceramic knives and scissors (as well as the largely unnoticed electronic substrate field) prior to their introduction in specialist engine/turbo charger applications or in the case of high Tc materials into antennae, SQUIDS etc.

Although the focus of this conference has been firmly on new "molecularly engineered" conducting polymers we should not forget that many technology "pull through" applications for these materials will be in enabling a new generation of "conducting compounds" (eg composites, paints, inks, pastes etc) to be fabricated where control of conductivity in the important 10^{-5} - $10\,\Omega$ cms range, going beyond the percolation limitations (and colour limitations) of carbon black become very important largely in commodity type applications. Such applications will be important in providing economies of scale as well as deriving benefit from the leading edge applications learning that will come from the areas where molecular engineering of intrinsic properties is the dominant driver.

2   CURRENT PROGNOSIS OF TECHNOLOGY IMPACT

The previous chapters illustrate the diversity of Technology applications which are currently the focus of research with likely impact in Electrical/Electronic, Automotive, IT, Aerospace and Process Industries. The "Engineered" effects include applications in sensors, displays, conductors, micro, molecular and bio devices, energy storage and in shielding and stealth.

An analysis of the unique properties which underpin these applications is given in Table 1.

TABLE 1 - NOVEL PROPERTIES OF CONDUCTING POLYMERS

* Ability to tune properties from same building blocks
* Ability to dope, implant
* Large dimensional changes in going from doped/undoped
* Ability to pattern, shape
* Ability to process in thin film form
* One D properties
* High and low frequency properties
* Surfaces vs bulk effects

The perceived Novel Technology dimensions are set out in Table 2 and we will return to perhaps the most interesting component of the properties/technology dimension associated with the intrinsic ability to integrate function in the next section.

TABLE 2 - NOVEL TECHNOLOGY

* Ability to fabricate and design in new shapes/patterns
* Wide range of possible engineering solutions to market needs

Aerospace, Defence	* Optoelectronics
Automotive	* Electronic
Electrical/Electronic	* Ionic
IT	* Electromagnetic in general

Process Industries

* Important theme of ability in principle to "Integrate function"

Table 3 summarises the Novel Physics aspects which have emerged from the research in this area in the past decade with the emphasis perhaps moving now towards how conducting polymers can uniquely provide the basis for the next generation of smart or intelligent materials systems which is the subject of the prognosis for the future.

TABLE 3 - NOVEL PHYSICS

* Band Gap Engineering
* New theoretical concepts
* New regime for computational modelling
* New devices
* One D properties
* Important potential component in "smart" materials design
* Low IQ <------> high IQ
  Sensing and feedback

3   PROGNOSIS FOR THE FUTURE

We can perhaps start with a direct analogy with Nature's "Engineered Solutions" in the macromolecular arena involving as it does unentangled chains, unique control of dispersity, molecular recognition, the Composite and Surface Engineering paradigms etc. Nature's designs are optimum for their function and as such imply <u>anisotropic molecular design</u>. The mechanical, electrical, optical etc properties (which have been alluded to in the previous chapters) of conducting polymers offers considerable scope for such anisotropic, "man made" molecular design. In this connection we may note that Nature has had 10^9 years to optimise designs and is only required to work within narrow performance limits compared with many applications in the man made world. We may therefore consider conceptually what new "Molecular Design" opportunities might arise from "Conducting Polymers" where the exceptional electrical, optical (EM properties in general), mechanical properties already approaching the theoretical limit coupled with processing advantages make these outstanding candidate materials for engineered designs/solutions which Integrate functions. It seems likely therefore that this generic class of materials will form the basis for a new generation of High IQ Materials Systems.

By way of background it is perhaps worthwhile to briefly review materials trends as indicated in Figure 2.

MATERIALS TRENDS

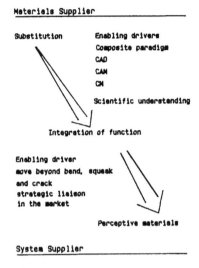

Figure 2

The prime focus of materials research in the polymer, compounds and composites field to date has been for "passive" applications many of which involve substitution of metals in existing applications the enabling drivers being the Composite paradigm (anisotropic design) and computer aided design, manufacture and modelling and the underpinning detailed molecular level understanding from advanced characterisation and measurement facilities. The future however lies in the ability to integrate many functions in "active" or "perceptive" materials since this will enable the move to be made from substitution to entirely new materials applications.

Since this emerging area will be new to many people we provide here a brief definition of the terminology.

An <u>intelligent</u> material/system is one capable of understanding and of obtaining information.

A <u>perceptive</u> material/system is one capable of intuitive recognition or action by which sensations to external signals/stimuli may be interpreted and acted on.

At a fundamental level a <u>low IQ</u> material/system is one which can sense its situation and provide information on its status whereas a <u>high IQ</u> material/system in addition to being able to report can also act and modify its properties commensurate with the needs of its environment. This is perhaps clearer in Figure 3 where we may envisage unique attributes of conducting polymers as being a key enabler in moving beyond the metals displacement mindset with passive materials into the arena of the "active" materials systems which will be increasingly important for the future.

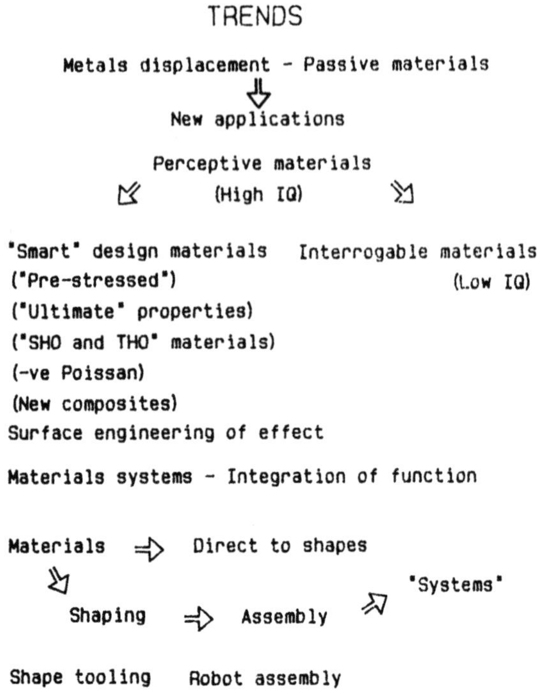

TRENDS

Metals displacement - Passive materials

New applications

Perceptive materials
(High IQ)

"Smart" design materials   Interrogable materials
("Pre-stressed")   (Low IQ)
("Ultimate" properties)
("SHO and THO" materials)
(-ve Poissan)
(New composites)
Surface engineering of effect

Materials systems - Integration of function

Materials ⟹ Direct to shapes
                                                    "Systems"
        Shaping ⟹ Assembly

Shape tooling   Robot assembly

Figure 3

As exemplars of this genre we consider briefly the potential generic application of the ability to integrate function which conducting polymers might have in the advanced structural composites field.

In looking at the role which may well be played by conducting polymers we can envisage that as well as forming part of the structural integrity of the system, the advanced sensor, the parallel information processing, the pattern recognition (spatial light modulator) and the feedback actuator (eg shape memory polymer system and or nanoscale motor) might all involve conducting polymers and although research is still in its infancy individual components of the technologies which will be required have been demonstrated in a number of areas.

Certainly in the aerospace arena self damping space platforms and "the feeling plane" with ability to make automatic corrections of flying surfaces are longer term targets that are likely to be impacted by developments in the conducting polymer field.

In terms of smart or intelligent structure and function we may generically consider 3 areas of application.

i)   Improvements in Manufacturing Process

    The ability to interrogate a materials system in real time during the shaping, fabrication, consolidation regimes is an important one which in the case of, for example high performance composites, requires not only real time sensors for data capture but also signal processing capability to facilitate process control loops to modify/optimise processing conditions to provide a complete history of the fabrication of a composite component. Conceptually one can already see how conducting polymer fibres incorporated into the structure could be used in a more advanced and complete way than the current generation of macroscopic glass optical fibres which are coming into application as first generation passive data capture approaches to process history assessment.

ii) Structural Integrity and Safety

As well as exceptionally high specific properties compared with metals the ability to routinely design anisotropic structures by controlled fibre placement and the ability therefore to incorporate the means to interrogate and modify the properties of the structure (eg by incorporation of conducting polymer fibres) gives considerable flexibility for novel design applications which would be difficult to envisage with metal based systems. For advanced applications therefore where high temperatures are not an issue polymer composites materials look to have a more ready extension to the next generation of intelligent based materials than is the case for metal based systems.

The early learning on the fatigue behaviour of aluminium alloys in the aerospace arena and the frequent inspection regimes which are instituted in areas of materials application which place the public at risk suggests considerable scope for designing the next generation of materials where real time monitoring of structural parameters would ensure that the design envelope was not exceeded in use and that advanced warning would be given for any repair that might be necessary. In the longer term of course the requirement will be for automatic self repair much in the way that feedback loops operate in the biological sphere in the self repair/healing of bone fractures.

Again in terms of data capture, data reduction and interpretation and response one can conceptually see how the unique characteristics of conducting polymers might play an important role in the total engineered system.

iii) Higher Performance

Whilst 1 and 2 would have a considerable impact on existing advanced composite materials technology it is clear that conducting polymers could form the basis for the next generation of "smart" structural activators particularly if this was integrated with the ability to process information optically.

Optical computing looks to be many years away but elements of pattern recognition from the use of thin polymeric film, spatial light modulators and the ability to interface with conventional (Si, GaAs) electronics in hybrid devices are on a much shorter horizon and this is likely to pose some interesting choices for the design engineer/materials scientist. For a given application the choice between an "ultimate" properties but dumb system and "lower" properties but smart system (with an ability to self repair) may be an interesting one.

Lest we give the impression that the future of smart materials systems is only likely to be of interest to the esoteric end of aerospace it is conceptually quite easy to see how the learning which will undoubtedly come from the leading edge of technology and will be strongly impacted by developments in the conducting polymer field will feed forward into "commodity" applications in Transportation, Civil Engineering, Manufacturing and Consumer products and this is outlined in Figure 4.

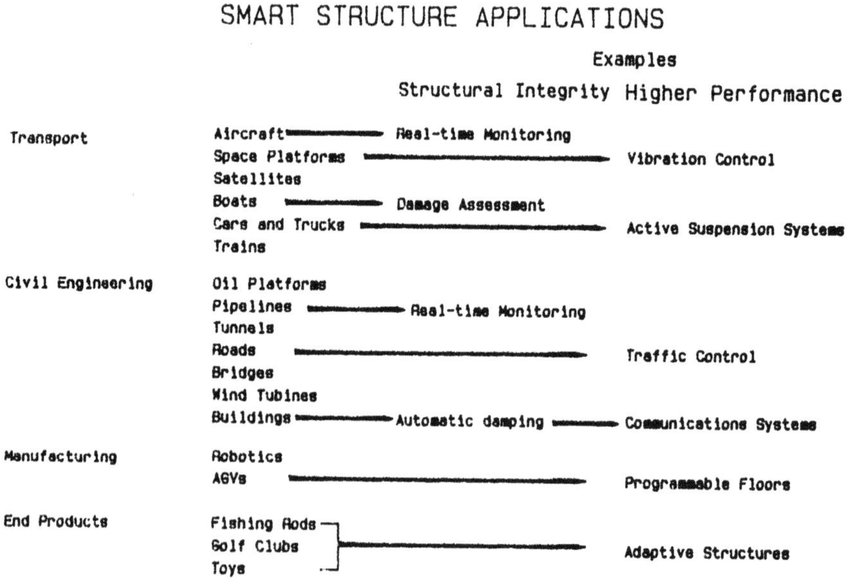

SMART STRUCTURE APPLICATIONS

Examples

Structural Integrity Higher Performance

Transport	Aircraft ⟶ Real-time Monitoring	
	Space Platforms ⟶	Vibration Control
	Satellites	
	Boats ⟶ Damage Assessment	
	Cars and Trucks ⟶	Active Suspension Systems
	Trains	
Civil Engineering	Oil Platforms	
	Pipelines ⟶ Real-time Monitoring	
	Tunnels	
	Roads ⟶	Traffic Control
	Bridges	
	Wind Tubines	
	Buildings ⟶ Automatic damping ⟶	Communications Systems
Manufacturing	Robotics	
	AGVs ⟶	Programmable Floors
End Products	Fishing Rods ⟶	
	Golf Clubs ⟶	Adaptive Structures
	Toys ⟶	

Figure 4

The golf clubs which adapt in real time to peoples preponderance to slice or undercut or undershoot their puts may be a half a generation away but conducting polymers could be the key to a development that takes the average golfer closer to the calibre of a Nick Faldo on a foreseeable horizon. The field is moving fast the preceding chapters have set out some of the scientific underpinning that will be the springboard for many innovative materials applications over the coming decade. The field is therefore one of timeliness and promise and we look forward to many of the applications that people currently envisage (and many more) coming to fruition over the next decade. One thing that is certain is that the opportunities provided by Conducting Polymers in engineering effects in shapes will help to drive the closer integration of the activities of the synthetic polymer chemists, the physicists measuring properties and the processing design and fabrication scientists. Indeed the various chapters of this book should emphasise that already the move from the LHS to the RHS of Figure 5 is already well under way in this field.

MATERIALS TRENDS

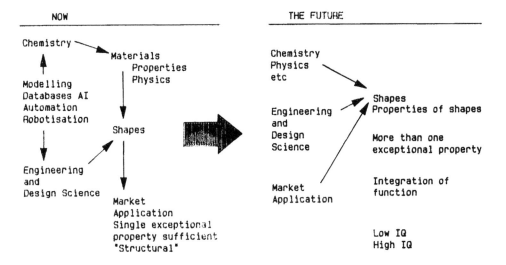

Figure 5

REFERENCES

1   J E Österholm Symposium presentation

2   A McDiarmid Symposium presentation and chapter

3   A Epstein communicated at the symposium

Electroactivity of poly(3-methyl-thiophene) in aqueous solutions of inorganic salts

Svein Sunde[1], Georg Hagen[1] and Rolf Ødegård[2]

1) Laboratories of Industrial Electrochemistry, Norwegian Institute of Technology, N-7034 Trondheim, Norway
2) SINTEF/Div. of Metallurgy, N-7034 Trondheim, Norway

Poly-(3-methyl-thiophene) (P3MeT) shows unusual strong sensitivity towards certain ions in aqueous solutions, unknown in organic solvents. Voltammograms of P3MeT in 0.1 M KNO_3(aq) exhibits a strong virginal anodic peak near 900 mV (Ag/AgCl), while the corresponding cathodic "peak" is smeared out over several hundred mV. The anodic current response of P3MeT is strong in NO_3^-(aq) and in ClO_4^-(aq) solutions, and weak in Cl^-(aq) and SO_4^{2-}(aq) solutions, which may indicate a correlation with the oxidation power of the anions employed. The strong response towards nitrate ions suggests the use of P3MeT as a sensor material.

The appearance of the voltammograms of P3MeT in aqueous solutions are dependent on the history of the electrode. During cycling in aqueous nitrate solutions the anodic peak potential is shifted strongly in the negative direction, and the peak current is significantly reduced. The anodic current can however be partly reactivated by dedoping again in acetonitile. Storing the polymer film electrode (e.g. for one week) in KNO_3(aq) results in similar voltammograms, but with a strongly enhanced current response, indicating the time dependence for electrolyte penetration into the porous polymer film.

Extending the vertex potentials in the anodic direction, reveals a second anodic peak, which is associated with degradation of the polymer, resulting in loss of electroactivity. The degradation peak is also observed to be solvent dependent, and is shifted towards more negative potentials in water compared to acetonitrile solutions. In aqueous solutions of Cl^- and SO_4^{2-} it is difficult to resolve the two anodic (doping and degradation) peaks by voltammetry. However, the doping process is observed by means of in-situ conductivity measurements.

Acknowledgement: This work was supported by the norwegian research council NAVF.

Oxidation factor for polyaniline films on ITO from cyclic voltammetry and visible absorption spectra

B.P. Jelle[x], G. Hagen[x], S.M. Hesjevik[+] and R. Ødegård[+]
[x] Laboratories of Industrial Electrochemistry
[+] SINTEF Div. of Metallurgy
N–7034 NTH Trondheim

Polyaniline (PANI) films have been deposited electrochemically on glass plates coated with indium–tin–oxide (ITO) from aqueous solutions of aniline in sulphuric acid. A simple formula for polyaniline is:

Reduced unit   Oxidized unit

where

y = oxidation factor for PANI $\in [0,1]$
x determines the length of the polymer chain

We have calculated the oxidation factor (y) for PANI as a function of potential from cyclic voltammograms and from visible absorption spectra.

We assume that the oxidation factor (y) is proportional to the charge (q) which has passed through the film during a potential sweep:

$$y = kq$$

where k is a constant. The charge was determined from the voltammograms, and assuming $y=1$ (fully oxidized) when $q=q_{max}$, y may be calculated.

Assuming that the observed absorption spectra for PANI films may be composed by superposition of the two individual spectra, for the fully oxidized form $(y=1)$ and the fully reduced form $(y=0)$, the absorption $C(y)$ may be expressed as:

$$C(y) = Ay + (1-y)B$$

where A and B are the absorptions for the fully oxidized and fully reduced form of PANI respectively. A certain wavelength is picked and by assuming $C(y=1)=C_{max}=A$ and $C(y=0)=C_{min}=B$, y may be calculated.

The oxidation factor (y) for PANI films calculated from cyclic voltammograms shows generally good agreement with calculations of y from visible absorption spectra of PANI.

This work is supported by The Royal Norwegian Council for Scientific and Industrial Research (NTNF).

Layer-by-layer chemical deposition of conducting polymer thin films

Y.F. Nicolau, S. Davied and M. Nechtschein
DRF/SPh/Dynamique de Spin et Propriétés Electroniques,
Centre d'Etudes Nucléaires de Grenoble, 85 X,
38041 Grenoble Cédex, France

We are studying a new method, based on the chemical oxidative poly-condensation, enabling the simultaneous synthesis and deposition of insoluble conducting polymer thin films like : PPY, PANI, PBT and P3MT. We obtain the films layer-by-layer (LL) on insulator, semiconductor or conductor substrates by successive and alternate soakings of the substrates in a monomer solution and in an oxidant solution. The films are LL cyclically synthesized in the liquid diffusion layer adherent to the polymer (substrate) - solution interface. The solid or liquid monomer is supplied as thick film adherent to the interface and the oxidant solution is renewed after each cycle. So, the deposition of individual chains and the lengthening of the already deposited ones found at the interface are enhanced and the deposition of agglomerates of globules is depressed. In order to obtain highly conducting films and to avoid over-oxidation and covalent oxygen bonding different concentrations of $FeCl_3$ - solvent oxidizing solutions are specially chosen. The deposition of oligomers is strongly depressed.

All films, from 40 to 1000 nm in thickness, are dense, show a granular morphology and an average roughness ranging from 10 to 100 nm, increasing with thickness. All films are crystallized, PPY excepted. The growth rate can be conveniently varied between 2 and 20 nm/cycle corresponding to 0.05-0.25 μm/h. The films of PPY, PANI, PBT and P3MT have a conductivity of 20-100, 3-15, 1-5 and 40-100 S/cm respectively. Till now measured conducting PANI and P3MT have a very low spin number : 1 spin/10^3-10^4 rings. As compared with electrochemically synthesized ones, LL films show : (i) less structural dissorder, (ii) a better crystallinity, (iii) a higher density and (iv) an equivalent (or higher) conjugation length. The films are suitable for making electronic or optical devices.

Infrared ellipsometry of thin, oriented polymer films

J. Bremer, O. Hunderi, F. Kong and T. Skauli

Department of Physics
The Norwegian Institute of Technology
N-7034 Trondheim - Norway

The recently developed technique of ellipsometry in the
infrared region of the electromagnetic spectrum provides
information about amplitude, phase and polarization state for
low-energy radiation reflected off a surface. We have built
and tested an infrared ellipsometer for the study of ultrathin
layers of organic/inorganic materials. Measurements have so
far been performed on doped GaAs films, electro-polymerized
polythiophene and polypyrrole, aerogels, various superlattices
and bulk inorganic crystals. The actual build-up is based on
a Fourier-transform spectrometer which has been equipped with
a computerized ellipsometric attachment. Positioning of the
polarizers and calibration of their azimuthal settings are
done automatically by a computer. Since oriented polymer
films generally behave as biaxial optical systems evaluation
of the data requires some special considerations. Furthermore, the sharp variation in the refractive index near
resonances may 'distort' the band shape of vibrational
reflection spectra. In addition, spectra from thicker films
exhibit the usual interference features. Comparison between
measured and calculated amplitudes/phases allows valuable
informations to be extracted. For example, in the case of
electropolymerized polythiophene in benzonitrile at 15 V
(J. Mårdalen et al., (1990)) the reflected s- and p-amplitudes
are found to behave strikingly different. In the p-case,
where the electric field vector has a non-vanishing component
in a direction perpendicularly to the surface of the Pt
substrate the aromatic CH-mode near 3000 cm^{-1} is found to be
absent. The vanishing mode indicates that an ordering
mechanism is present during film-growth, thereby favouring a
structure where the thiophene rings are parallel to the
substrate surface. This conclusion is in qualitative
agreement with the x-ray recordings of Ito et al. (Journal of
Polymer Science, C 24, 147-151 (1986)). Infrared ellipsometry
combines the multiplex advantage of Fourier-transform
spectroscopy and the phase sensitivity of ellipsometry. A
full polarometric analysis is necessary when interpreting data
from a sample with a film-like structure.

Aqueous application technology of conductive layers for photographic materials

W.De Winter   Agfa N.V., R&D Department

Ever since they have been developed, photographic films require the use of antistatic outermost layers in order to prevent dust-uptake and to allow for (fast) transportation in cameras and processing machines without causing electrostatic discharges.

A common procedure to realize these antistatic layers consists in the use of polymeric binding agents containing ionizable groups . The electrical properties of such layers, however, may vary with changes in the degree of humidity (RH) of the environment. Attempts to diminish the RH-sensitivity by 'strengthening' the layer structure through cross-linking (via the 'Conducto-Cross'-technique or by using cross-linked latex), did not result in the improvements looked for; particularly with regard to applicability in electronic imaging techniques.

The development of films of polymers with conjugated - or charge-transfer-systems has opened the way towards RH-independent, highly conductive layers, which at the start, however, were also highly colored, making them unsuited for photographic applications. Announcements of new, transparent, colorless polymers (e.g. polyisothianaphthene), applicable from organic solvents, show great promise. Practice, however, has not yet matched theory. In addition, ecological legislation becomes tougher every day, so that it is desirable to make non-toxic conductive layers from aqueous media rather than from organic ones. Attempts to prepare conductive layers from polymers with (doped) conjugated systems from an aqueous medium have been carried out via two routes : by post-treatment of water-soluble precursor polymers, and by applying the 'loaded-latex'-technique. Both techniques have been successful to some degree, but no completely satisfactory "systems" have been realised so far.

Therefore, quite a research effort will be needed in order to meet all the requirements with regard to properties and application technology for future (photographic) imaging materials.

Chemically prepared polyalkylthiopenes, polydialkylbithiophenes and polyalkoxythiophenes: Spectroscopic and electrochemical studies

M.Zagórska, I.Kulszewicz-Bajer, M.Hasik*, J.Laska*, A.Proń*

Department of Chemistry, Technical University of Warsaw
00-664 Warszawa, Noakowskiego 3 (Poland)

*Department of Materials Science and Ceramics, Academy of
Mining and Metallurgy, Kraków, al.Mickiewicza 30 (Poland)

Three types of compounds from polythiopene family of conducting polymers have been studied: poly(3-alkylthiophenes),
poly(4,4´-dialkylbithiophenes) and poly(3-alkoxythiophenes).
In each case $FeCl_3$ was used as the oxidizing/polymerizing
agent. Polydialkylbithiophenes distinguish themselves from
the other two groups by their better stereoregularity which
is manifested in ^{1}H NMR and cyclic voltammetry studies. This
observation is further corroborated by optical and spin responses to the potential change which show that polaron formation and their anihilation to bipolarons occurs in a much
narrower potential range as compared to polyalkylthiophenes
and polyalkoxythiophenes.

Electropolymerization of thiophene and substituted thiophenes on ITO-glass by high voltage method

J.Mårdalen[1], E.J.Samuelsen[1], E. Olsen[1], O.R.Gautun[2], P.H.Carlsen[2], S.Sunde[3]

[1]Institute of Physics, [2]Institute of Organic Chemistry, and [3]Institute of Industrial Electrochemistry, The Norwegian Institute of Technology, N-7034 Trondheim-NTH, Norway.

Abstract
Films of polythiophene (PT), poly(3-metylthiophene) (PMT), poly(3-hexylthiophene) (PHT), and poly(3-octylthiophene) (POT) were electropolymerized on ITO-covered glass anodes under an applied voltage of 8V v.s. an Ag/AgCl reference electrode. Uniaxial stretching of polythiophene prepared in this way has previously been reported [1,2]. The films were electrochemically reversed, and further reduced by boiling in methanol to give deep red, flexible films with fairly smooth surfaces. The smoothness decreases with increasing length of the substitutents, and increasing film thickness. Films of PHT and POT are partially soluble in $CHCl_3$.

Fourier transform infrared (FTIR) spectra of the substituted polythiophenes are in agreement with previously reported data [3]. Cyclic voltammetry shows further that the films are electroactive. It was possible to cycle more than 10 times. This indicates that overoxidation of the films during polymerization is not severe. Doping with saturated I_2 in CCl_4 gave conductivity of about 10 S/cm.

By X-ray diffraction films of PMT was found to be amorphous and in agreement with diffraction profiles reported by Winokur et al. [4]. High voltage electrochemically prepared PHT and POT on the other hand were found to be more crystalline than chemically prepared samples. This is in contrast with previous reports [4].

UPS spectra of PHT show a π-band structure comparable to those of chemically prepared PHT [5]. Optical absorption show further a 2 eV band gap which is in full agreement with previous reported data. Thermochromic effect observed for PHT and POT is in agreement with results presented by Salaneck et al. [6].

References
1 M.Satoh, H.Yamasaki, S.Aoki, K.Yoshino, *Polym. Commun.*, **28** 144 (1987)
2 M.Satoh, H.Yamasaki, S.Aoki, K.Yoshino, *Mol.Cryst.Liq.Cryst.*, **159** 289 (1988)
3 S.Hotta, W.Shimotsuma, M.Taketani *Synthetic Metals,* **10** 85 (1984/85)
4 M.J.Winokur, D.Spiegel, Y.Kim, S.Hotta, A.J.Heeger, *Synth. Met.*, **28** C419 (1989)
5 M.Löglund, R.Lazzaroni, S.Stafström, W.R.Salaneck, J.-L.Bredas, *Phys. Rev. Lett.,* **63** 1841 (1989)
6 W.R.Salaneck, O.Inganäs, B.Themans, J.O.Nilsson, B.Sjögren, J.-E.Österholm, J.-L.Bredas, S.Svensson, *J.Chem.Phys.,* **89** 4613 (1988)

Acknowledgement
The authors are greatly indebted to W.R.Salaneck and M.Löglund, Linköping for measuring UPS and optical absorption. Financial support from Norges Allmenvitenskaplige Forskningsråd (NAVF) through the materials research programme is acknowledged.

Conducting polymers based on triazoles?

Per H. Carlsen[1], Odd Gautun[1], Erik Høgh Iversen[1], Jostein Mårdalen[2], Emil J. Samuelsen[2], Göran Helgesson[3] and Susan Jagner[3].

[1] Institutt for organiske kjemi og [2] Institutt for fysikk, Universitetet i Trondheim - Norges tekniske høgskole, N-7034 TRONDHEIM-NTH, [3] Institutionen för oorganisk kemi, Chalmers tekniske högskola, S-412 96 GÖTEBORG.

The molecular and crystallographic structures of some diphenyltriazoles

and some related materials are being investigated with a view of electron delocalization and possible conducting polymer formation.

(a) with R_1 = ethyl crystallizes in the polar space group C_s^4 with ring twist angles $44°$ and $48°$, whereas (b) with R_2 = propyl crystallizes in $P2_1/c$ with $49°$ and $53°$ twists. (b) with R_2 = H is $P2_1/n$ with smaller twists, $8°$ and $9°$. The C-ϕ distance is in all cases 1.47 - 1.48 Å, all indicating only modest electron conjugation. Electropolymerization has not been successful. Preliminary data has been obtained of (a) with R_1 = dimethyltriazole.

Work is in progress on the structure of di-thiophene-dicarbylamine

whose crystal structure is pseudo-orthorhombic with partially disordered groups.

Subject Index

T - #0227 - 111024 - C0 - 234/156/9 - PB - 9780367403003 - Gloss Lamination